Pragmatic Flutter

Pragmatic Flutter

Building Cross-Platform Mobile Apps for Android, iOS, Web, & Desktop

Priyanka Tyagi

CRC Press
Taylor & Francis Group
Boca Raton London New York

CRC Press is an imprint of the
Taylor & Francis Group, an **informa** business

First edition published 2022
by CRC Press
6000 Broken Sound Parkway NW, Suite 300, Boca Raton, FL 33487-2742

and by CRC Press
2 Park Square, Milton Park, Abingdon, Oxon, OX14 4RN

© 2022 Taylor & Francis Group, LLC

CRC Press is an imprint of Taylor & Francis Group, LLC

ISBN: 978-0-367-61209-2 (hbk)
ISBN: 978-1-032-05565-7 (pbk)
ISBN: 978-1-003-10463-6 (ebk)

Typeset in Times LT Std
by KnowledgeWorks Global Ltd.

To my husband Krishna
And my children Kalp and Krisha
Who inspired me to start writing this book and finish it!

Contents

Preface

Have you ever thought of creating beautiful blazing-fast native apps for iOS and Android from a single codebase? Have you dreamt of taking your native apps to the web and desktop without costing a fortune? If so, this book is the right place to start your journey of developing cross-platform apps.

Google's Flutter software development kit (SDK) is the latest way of developing beautiful, fluid cross-platform apps for Android, iOS, Web, and Desktops (macOS, Linux, Windows). Google's new Fuchsia OS user interface (UI) is implemented using Flutter as well. Learning to develop mobile apps with Flutter opens the door to multiple devices, form-factors, and platforms from a single codebase.

You don't need any prior experience using Dart to follow along with this book. However, it's recommended to have some familiarity with writing code in one of the object-oriented programming languages like Java and Python. You will pick up the fundamentals of Dart 2 language in the first chapter. We will learn to structure and organize the multi-platform Flutter project and test the setup by running the same code on Android, iOS, web, and desktop platforms. Next, we will explore basic Flutter widgets along with layout widgets while building a non-trivial UI. Later on, we will continue learning to build responsive layouts for different screen sizes and form-factors.

We will be organizing and applying themes and styles, handling user input, and gestures. Then we will move on to fetching data over the network, integrating, and consuming REST API in app. We will build a *BooksApp* to display books listing from Google Books API and integrate this API to fetch book data. Different types of testing strategies will be introduced to develop solid and high-quality apps. You will get hands-on experience using state management solutions, data modeling, routing, and navigation for multi-screen apps. You will learn to use Flutter plugins to persist data in the app. This book concludes by giving the pointers to deploy and distribute your Flutter app across all four platforms.

When you finish this book, you will have a solid foundational knowledge of Flutter SDK that will help you move forward in your journey of building great and successful mobile apps that can be deployed to Android, iOS, web, and desktop (macOS) from a single code base.

Throughout the book, *italic* is used to render file, folder, and Flutter application names. The monospace font is used to represent code snippets, variable names, and data structures. All examples used in this book are available online in the GitHub repository: https://github.com/ptyagicodecamp/pragmatic_flutter.

Last but definitely not least, I want to acknowledge the input of my fellow developers Rajalakshmi Balaji, Preetika Tyagi, and Snehal Patil without whom this book wouldn't be possible. I am also thankful to the Flutter community worldwide for sharing their expertise and knowledge for developing great cross-platform applications.

Author

Priyanka Tyagi is a Software Engineering Manager at Willow Innovations, Inc., which builds products that improve the lives and health of women. At Willow Innovations, Inc., she is helping to build a better wearables experience by using Flutter and Bluetooth Low Energy (BLE) technologies.

Priyanka has many years of experience designing and developing software, web, and mobile systems for a diverse range of industries from automobile and e-commerce to entertainment and EdTech to health and wellness. Her expertise lies in Flutter, Android, Firebase, Mobile SDKs, AWS/Google cloud-based solutions, cross-platform apps, and game-based learning. In her previous roles, she has worked with Disney Interactive as Lead Android Engineer. She has also helped various start-ups with her software engineering consulting services.

Priyanka loves to share her tech explorations around mobile apps development and other software engineering topics in her tech blog: https://priyankatyagi.dev/.

She is an Internet of Things (IOT) enthusiast and volunteers her time at local public schools to introduce computer science to young minds. She volunteers for the Hour of Code initiative to inspire young developers in the tech world for many years. Priyanka is passionate about mentoring aspiring developers to get started in the tech industry. She earned an MS in computer science from Illinois Institute of Technology, Chicago.

Priyanka lives with her husband and two kids in beautiful California. She loves to read, bake, and hike in her free time.

1 Dart Fundamentals
A Quick Reference to Dart 2

Dart programming language is developed by Google. It has been around since 2011. However, it has gained popularity recently since Google announced Flutter software development kit (SDK) for developing cross-platform-applications. Dart aims to help developers build web and mobile applications effectively. It works for building production-ready solutions for client applications as well as for the server-side. Dart has an Ahead-of-Time (AOT) compiler that compiles predictable native code quickly for the target platform. It is optimized to build customized user interfaces natively for multiple platforms. The Dart is a developers-friendly programming language. It is easy for developers coming from different programming language backgrounds to learn Dart without much effort.

The Dart 2 (Announcing Dart 2 Stable and the Dart Web Platform) is the latest version of Dart language including the rewrite of many features, improved performance and productivity. This chapter covers the basics of Dart 2 language syntaxes (Google, 2020) to get started with the journey of building applications using Flutter. A prior understanding of object-oriented programming is needed to follow along with the material. Dart 2 and Dart will be used to infer the Dart 2 programming language in this book. In the upcoming topics, we'll review some of the language features with the help of examples.

THE main FUNCTION

The `main()` function is the entry point for any Dart program. The following code snippet will print "Hello Dart" on the console. Open a 'Terminal' at MacOS or its Windows or Linux equivalent. Create a file say '*hello.dart*', and copy the following code snippet in it.

```
void main() {
     print("Hello Dart");
}
```

RUNNING DART PROGRAM

Go back to the 'Terminal' and execute the file using `dart hello.dart`. The "Hello Dart" is printed on the console.

```
$ dart hello.dart
Hello Dart
```

VARIABLES & DATA TYPES

In Dart, you can use the `var` keyword to infer the underlying data type. The following code snippet shows declaring variable `data` to hold the numeric value '1'.

```
var data = 1;
print(data);
```

The above code snippet will print the '1' on the console. The runtimeType (runtimeType property) property gives the runtime type of the object it's called on. The following code snippet will print the data type of the `data` variable as 'int' on the console.

```
var data = 1;
print(data.runtimeType);
```

It's valid for the `data` variable to get reassigned to a value from a different data type.

There's another keyword, `dynamic`, to assign a value to a variable similar to the `var` keyword. The difference between the two is that for the `dynamic` keyword, the same variable can be reassigned to a different data type, as shown in the code snippet below:

```
// Assigning int to dynamicData
dynamic dynamicData = 1;
// Prints `int` as data type on console
print(dynamicData.runtimeType);

//Re-assigning dynamicData to String data type
dynamicData = "I'm a string now";
// Prints `String` as data type on console
print(dynamicData.runtimeType);
```

SOURCE CODE ONLINE

The source code for this example (Chapter01: Quick Reference to Dart2 (Variables & Data Types)) is available at GitHub.

COLLECTIONS

LIST

Dart's List (List<E> class) is a collection that holds indexable objects. An empty list can is declared using two square brackets '[]'. List items can be put inside these brackets separated by commas.

```
List emptyList = [];
```

```
//Checking if List is empty
var result = emptyList.isEmpty;
print(result);
```

SPREAD OPERATOR

Dart 2.3 introduced the spread operator (Spread Operator) and null aware spread operator to combine multiple lists into one. Let's create a list `theList` with two items and a null list `theNullList`. If you want to merge these two lists into another list, say `anotherList`, it can be done using spread (...) and null aware spread (?...) operators.

```
List theList = ['Dart', 'Kotlin'];
//null list
List theNullList;

// Spread operator ... - Flatten the list and to be merged
with another list
// Null spread operator ...? - Helps to avoid runtime
exceptions - iterator called on null.
List anotherList = ['Java', ...theList, ...?theNullList];

// Printing the merged list created using spread operators
print(anotherList);

// Output
[Java, Dart, Kotlin]
```

TRANSFORM LIST ITEMS

The `map` on the `List` can be called to transform the list items. Let's append the word 'Language' to objects of the `theList` list.

```
List theList = ['Dart', 'Kotlin'];
result = theList.map((e) => "$e Language").toList();
print(result);

// Output
[Dart Language, Kotlin Language]
```

FILTERING

The `where` method on `List` can help filter items meeting specific criteria. In the following code snippet, we are filtering the words containing the letter 'a'. Since only 'Dart' matches these criteria, `result` will hold only the word 'Dart'.

```
```
List theList = ['Dart', 'Kotlin'];
result = theList.where((element) =>
element.toString().contains('a'));
print(result); // Dart contains letter 'a'

// Output
(Dart)
```
```

SET

The Set (Set<E> class) data structure holds a collection of objects only once. The duplicates are not allowed when storing data in `Set`. Let's create two sets to learn the usage.

```
```
Set langSet = {'Dart', 'Kotlin', 'Swift'};
Set sdkSet = {'Flutter', 'Android', 'iOS'};
```
```

ADDING ITEM

Let's add a new item 'Java' in the `langSet` and print the final set.

```
```
// Adding 'Java' to Set
langSet.add('Java');
print(langSet);

// Output
{Dart, Kotlin, Swift, Java}
```
```

REMOVING ITEM

Let's remove the newly added item 'Java' from `langSet` and print the final set.

```
```
// Remove Java from Set
langSet.remove('Java');
print(langSet);

// Output
{Dart, Kotlin, Swift}
```
```

ADDING MULTIPLE ITEMS

More than one item can be added all at once in `Set` using the `addAll()` method. It accepts the list of items. Let's add multiple items to both sets.

```
```
// Adding multiple items to each set
langSet.addAll(['C#', 'Java']);
sdkSet.addAll(['C#', 'Xamarin']);
print(langSet);
print(sdkSet);

// Output
{Dart, Kotlin, Swift, C#, Java}
{Flutter, Android, iOS, C#, Xamarin}
```
```

INTERSECTION OF TWO SET(S)

The `intersection` method on `Set` returns the shared items from both sets. The `langSet` and `sdkSet` both contain the 'C#' item in them. The intersection on those sets will return the 'C#'.

```
```
// Find Intersection of two sets (common items)
result = langSet.intersection(sdkSet);
print(result); // C#

// Output
{C#}
```
```

UNION OF TWO SET(S)

The `union` returns items combined from `langSet` and `sdkSet` without any duplicates. Both sets contain 'C#', and it'll be added in the `result` only once.

```
```
// Find a Union of two sets. No duplicates.
result = langSet.union(sdkSet);
print(result);

// Output
{Dart, Kotlin, Swift, C#, Java, Flutter, Android, iOS, Xamarin}
```
```

MAP

The Map (Map<K, V> class) data structure is a collection that contains key/value pairs. A value is accessed using the key for that entry. Let's create a `Map` with a key of `int` type, and value is of `String` type.

One way to create a Map `intToStringMap` is as below:

```
```
var intToStringMap = Map<int, String>();
```
```

The new key/value pair can be added as shown in the code snippet below:

```
intToStringMap[1] = '1';
intToStringMap[2] = '2';
```

The first or last entry of the map can be accessed using `first` and `last` properties of `entries` (entries property) iterable.

```
// first Map entry
result = intToStringMap.entries.first;
print(result);
// Output
MapEntry(1: 1)

// last Map entry
result = intToStringMap.entries.last;
print(result);
// Output
MapEntry(2: 2)
```

Let's create a `Map` with the key and value of type `String` as below:

```
var techMap = {
  'Flutter': 'Dart',
  'Android': 'Kotlin',
  'iOS': 'Swift',
};
```

CHECKING FOR KEY

The `containsKey(String key)` method on the map returns a boolean true or false depending on whether the given `key` exists in the Map or not.

```
// Returns boolean. true if key is found, otherwise false
result = techMap.containsKey('Flutter');
print(result);

// Output
true
```

CHECKING FOR VALUE

The `containsValue(String value)` method returns the boolean value as 'true' if the given value exists in the Map.

```
~~~
// Checking if value is present in the Map
result = techMap.containsValue('Dart');
print(result);

// Output
true
~~~
```

ACCESSING ALL VALUES

All values of the key/value pairs in Map collection can be accessed calling `foreach` on `values` as below:

```
~~~
// Prints all values
techMap.values.forEach((element) {
    print("$element");
});

// Output
Dart
Kotlin
Swift
~~~
```

ITERATING KEY/VALUE PAIRS

All key/value pairs in Map collection are iterated, as shown in the code below:

```
~~~
// Iterates over all key-value pairs and prints them
techMap.entries.forEach((element) {
    print("${element.value} is used for developing
    ${element.key} applications.");
});

// Output
Dart is used for developing Flutter applications.
Kotlin is used for creating Android applications.
Swift is used for developing iOS applications.
~~~
```

SOURCE CODE ONLINE

The source code for this example (Chapter01: Quick Reference to Dart2 (Collections)) is available at GitHub.

FUNCTIONS

Let's create a function that checks if the passed argument is 'Flutter' or not and returns a boolean value. It returns `true` if it's precise 'Flutter' or `false` otherwise. Such functions are known as 'Named Function' because the function's name describes what they are intended to.

```
```
bool isFlutter(String str) {
    return str == 'Flutter';
}
// Using Named Function
dynamic result = isFlutter("Flutter");
print(result);

//Output
true
```
```

FUNCTION WITH OPTIONAL PARAMETERS

Let's create a function `concat(...)` that joins two strings together when the second string is available. In such a case, the second string can be passed as optional using square brackets '[]'.

```
```
String concat(String str1, [String str2]) {
    return str2 != null ? "$str1 $str2" : str1;
}

// Usage
result = concat("Priyanka", "Tyagi");
print(result);

//Output
Priyanka Tyagi
```
```

NAMED PARAMETERS

The other way to provide optional parameters is to use named parameters. They can be passed using curly braces '{}'.

```
```
// Named Parameters: Function with optional parameters in
curly braces
String concat2(String str1, {String str2}) {
    return str2 != null ? "$str1 $str2" : str1;
}
```
```

```
// Using function with optional params with curly braces{}
result = concat2("Priyanka", str2: "Tyagi");
print(result);
// Output
Priyanka Tyagi
```
```
```

PASSING FUNCTION AS PARAMETER

The Dart 2 programming language allows passing a function as a parameter to another function. Let's create a function `int subtract(int a, int b)` to find the difference between the two values 'a' and 'b'. The method `calculate(...)` takes two numbers and a function as a parameter to operate on the values passed.

```
```
```
int subtract(int a, int b) {
    return a > b ? a - b : b - a;
}
// Passing Function as parameter
int calculate(int value1, int value2, Function(int, int)
function) {
    return function(value1, value2);
}
//Passing function 'subtract' as parameter
result = calculate(5, 4, subtract);
print(result);

// Output
1
```
```
```

ARROW SYNTAX

The arrow syntax can be used to write functions in one line. Let's write the function to add two numbers using arrow syntax as below:

```
```
```
// Arrow Syntax
int add(int a, int b) => a + b;

// Using Arrow Syntax
result = add(5, 4);
print(result);

//Output
9
```
```
```

Anonymous Function

Anonymous functions don't have a name and can be assigned to a variable either using the keyword `var` or `Function`.

```
~ ~ ~
// Anonymous Function
Function anonymousAdd = (int a, int b) {
    return a + b;
};

// Using anonymous functions.
// Calling function variable `anonymousAdd`
result = anonymousAdd(4, 5);
print(result);

//Output
9
~ ~ ~
```

Source Code Online

The source code for this example (Chapter01: Quick Reference to Dart2 (Functions)) is available at GitHub.

CLASSES

Dart classes are created using the keyword `class`. Let's define a class Person to represent a person in the real world. This person has a name, an age, and the food it eats.

```
~ ~ ~
class Person {
    String name;
    int age;
    String food;
}
~ ~ ~
```

Constructor

Dart supports easy to use constructors. Let's see two types of constructors. The short-form constructor looks like below. The first part is required. The parameters inside '[]' are optional.

```
~ ~ ~
Person(this.name, [this.age]);

//Usage
Person person = Person("Priyanka");
~ ~ ~
```

Another type of constructor is the named constructor. All parameters are enclosed in the curly braces '{}'.

```
```
Person.basicInfo({this.name, this.age});

//Usage
Person child = Person.basicInfo(name: "Krisha", age: 6);
```
```

GETTERS

The getters in Dart classes are defined using the `get` keyword. Let's create a getter `personName` to get the name.

```
```
String get personName => this.name;
```
```

SETTERS

The setters in the Dart classes are defined using the `set` keyword. Let's create a setter `personName` to set the name as below:

```
```
set personName(String value) => this.name = value;
```
```

METHOD

Let's add a method to the class `Person` to define the eating behavior. The method `eats(String food)` takes the food as `String` and assigns it to the class `food` property.

```
```
void eats(String food) {
    this.food = food;
}
// Usage
Person child = Person.basicInfo(name: "Krisha", age: 6);
child.eats("Pizza");
```
```

Let's override the method `toString()` from the Object (Object class) class to print a custom message. Every class in Dart extends from the base Object class.

```
```
String toString() {
    return "My name is $name, and I like to eat $food";
}
```
```

```
//Usage
print(child.toString());

//Output on console
My name is Krisha, and I like to eat Pizza
```

CASCADING SYNTAX

Dart supports cascading syntaxes. It's useful in assigning the values to properties and methods at once using two dots.

```
child
 . .name = 'Kalp'
 . .eats("Pasta");
```

The setters can also be called using cascaded syntax as shown in the code snippet below:

```
child
 . .personName = 'Tanmay'
 . .eats("Pesto");
```

SOURCE CODE ONLINE

The source code for this example (Chapter01: Quick Reference to Dart2 (Classes)) is available at GitHub.

CONCLUSION

In this chapter, we reviewed the basic concepts and syntaxes useful to getting started with developing Flutter application quickly and efficiently. You were introduced to the structure of the Dart program. You developed your understanding of declaring and using variables in Dart language. We explored methods to store data in collections using various data structures like List, Map, and Set. You also learned to write reusable code using functions. Finally, you learned to architect and structure the code base using Classes and methods.

REFERENCES

Dart Dev. (2020, 12 14). *entries property*. Retrieved from api.dart.dev: https://api.dart.dev/stable/2.10.3/dart-core/Map/entries.html
Google. (2020, 11 13). *A tour of the Dart language*. Retrieved from Official Dart Documentation: https://dart.dev/guides/language/language-tour

Google. (2020, 11 13). *List<E> class*. Retrieved from Official Dart Language Documentation: https://api.dart.dev/stable/2.10.3/dart-core/List-class.html

Google. (2020, 11 13). *Map<K, V> class*. Retrieved from Official Dart Language Documentation: https://api.dart.dev/stable/2.10.3/dart-core/Map-class.html

Google. (2020, 11 13). *Object class*. Retrieved from Official Dart Language Documentation: https://api.dart.dev/stable/2.10.3/dart-core/Object-class.html

Google. (2020, 11 13). *runtimeType property*. Retrieved from Official Dart Documentation: https://api.dart.dev/stable/2.10.2/dart-core/Object/runtimeType.html

Google. (2020, 11 13). *Set<E> class*. Retrieved from Official Dart Language Documentation: https://api.dart.dev/stable/2.10.3/dart-core/Set-class.html

Google. (2020, 11 13). *Spread Operator*. Retrieved from Official Dart Language Documentation: https://dart.dev/guides/language/language-tour#spread-operator

Moore, K. (2018, 08 07). *Announcing Dart 2 Stable and the Dart Web Platform*. Retrieved from Medium: https://medium.com/dartlang/dart-2-stable-and-the-dart-web-platform-3775d5f8eac7

Tyagi, P. (2020, 11 13). *Chapter01: Quick Reference to Dart2 (Classes)*. Retrieved from Pragmatic Flutter GitHub Repo: https://github.com/ptyagicodecamp/pragmatic_flutter/blob/master/lib/chapter01/classes.dart

Tyagi, P. (2020, 11 13). *Chapter01: Quick Reference to Dart2 (Collections)*. Retrieved from Pragmatic Flutter GitHub Repo: https://github.com/ptyagicodecamp/pragmatic_flutter/blob/master/lib/chapter01/collections.dart

Tyagi, P. (2020, 11 13). *Chapter01: Quick Reference to Dart2 (Functions)*. Retrieved from Pragmatic Flutter GitHub Repo: https://github.com/ptyagicodecamp/pragmatic_flutter/blob/master/lib/chapter01/functions.dart

Tyagi, P. (2020, 11 13). *Chapter01: Quick Reference to Dart2 (Variables & Data Types)*. Retrieved from Pragmatic Flutter GitHub Repo: https://github.com/ptyagicodecamp/pragmatic_flutter/blob/master/lib/chapter01/variables.dart

2 Introduction to Flutter

Flutter is an open-sourced software development kit (SDK) (Flutter) to develop cross-platform applications and is free to use. It's Google's SDK for crafting beautiful, fast, and natively compiled applications for mobile, web, and desktop from a single code base. Flutter applications are written in Dart language (Dart). The Dart is optimized for building custom user interfaces (UIs) and fast cross-platform applications. The Dart code compiles to Native machine code for the platform app is running on. For example, the web Dart code compiles to JavaScript. Flutter is a complete framework that provides UI rendering & widgets, state management solutions, navigation, testing, and hardware application programming interfaces (APIs) to interact with device-level features like sensors, Bluetooth, etc.

CROSS-PLATFORM SOLUTIONS

The term 'cross-platform solutions' stands for the code reused across the different platforms or 'write code once and run everywhere'. The perfect cross-platform solution with a hundred percent code reuse is the holy grails that every mobile application developer dreams of. Let's discuss some of the solutions used to maximize the sharable code across multiple deployment targets like web, Android, iPhone Operating System (iOS), and desktop.

NATIVE SDKS

In the world of mobile app development, Native SDKs (Software development kit) prominently refer to Apple's iOS SDK (Develop iOS Applications) and Google's Android SDK (Develop Android Applications) to build mobile applications on iOS and Android platforms, respectively. The iOS applications are developed using either Objective-C (Objective-C) or Swift (Swift) programming languages. The Android applications are developed using Java (Java) or Kotlin (A modern programming language that makes developers happier.) programming languages. Your app runs code natively on the platform. The Native code is compiled to run on a particular platform like Android or iOS.

In this architecture, as shown in Figure 2.1, the application 'App-X' talks to the platform and accesses platform services directly. The platform side creates widgets that are rendered on the screen's canvas. The events are propagated back to the widgets. In this case, you would need to write two separate implementations of 'App-X' for each platform. Different implementations mean both apps are written in two entirely different languages as well.

Android and iOS platforms support legacy (Objective-C for iOS and Java for Android) and modern (Swift for iOS, and Kotlin for Android) programming

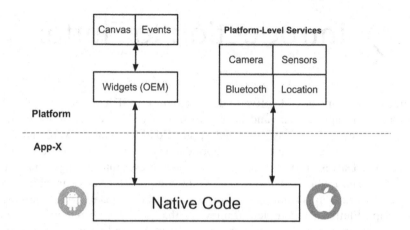

FIGURE 2.1   Native platform architecture

languages to develop applications. To create the same application for Android and iOS, a developer should be proficient in four programming languages all at once. Things become more complicated if legacy and modern languages are used together in the same code base. It may not be trivial to find developers who are experts in all four languages simultaneously. As you can see, such scenarios are not much cost-friendly. Since two separate applications need to keep the feature parity, developers need to implement the same feature twice from scratch that can take longer.

These are some of the issues that developers may want to avoid by using cross-platform solutions to develop applications that can share a single codebase in one programming language.

JavaScript + WebView

One solution for developing cross-platform applications from a single code is to use JavaScript with WebView(s). A WebView is a part of the browser's engine that can be inserted into a Native app along with loading web content in it.

In this architecture, as shown in Figure 2.2, the application 'App-X' creates hypertext markup language (HTML) contents and displays them in a WebView on the platform. The HTML/cascading style sheets (CSS) is used to recreate the Native original equipment manufacturer (OEM) widgets. The app is written using a combination of JavaScript, HTML, and CSS. The platform's services are accessed over the JavaScript bridge. The 'bridge' does the context switches between the JavaScript and the Native realms. It converts JavaScript instructions into Native instructions and vice versa.

Some of the cross-platform frameworks that are based on this approach are PhoneGap (Adobe) (Adobe PhoneGap), Apache Cordova, and Ionic. At the time of this writing, Adobe discontinued investment in PhoneGap and Cordova.

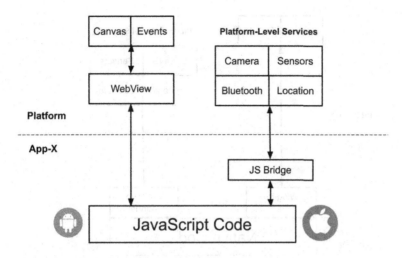

FIGURE 2.2 JavaScript and WebView architecture

REACTIVE NATIVE

In this approach, the creation of web views is simplified using design patterns based on reactive programming (Reactive Programming). React Native solution was developed by Facebook Inc. in 2005. It implements reactive style views on mobile applications.

In this architecture, as shown in Figure 2.3, the JavaScript code communicates with OEM widgets and platform services over the 'bridge'. This bridge does the context switching from the JavaScript realm to the Native realm and vice versa. This

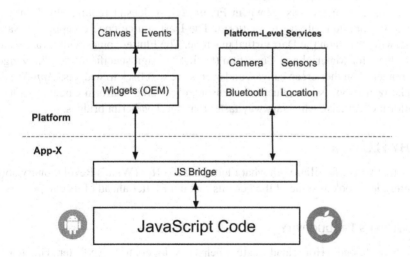

FIGURE 2.3 React Native architecture

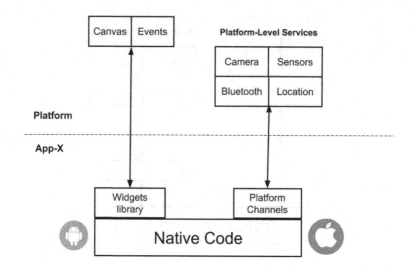

FIGURE 2.4   Flutter architecture

context switching happens for every widget and platform-service request that can cause performance issues and make sometimes widget rendering glitchy.

FLUTTER

The Flutter framework was developed by Google in 2017. Flutter is based on reactive-style views like React Native. It doesn't use bridges to talk to the platform and avoid performance issues caused by the context switching from the JavaScript realm to the Native realm. Flutter uses Dart language, which compiles ahead-of-time (AOT) into Native code for multiple platforms.

In this architecture, as shown in Figure 2.4, it doesn't require any JavaScript bridge to communicate to the platform. The Dart source code compiles to Native code, which runs on the Dart virtual machine. The Flutter comes with a vast library of widgets for Material and Cupertino (for iOS) design specifications. The widgets are rendered on the screen canvas, and events are sent back to widgets. 'App-X' talks to platform services like Bluetooth, sensors, etc., using Platform Channels (Flutter Platform Channels), which are way faster than the JavaScript bridges.

WHY FLUTTER

Now that you're familiar with what Flutter has to offer (What's Revolutionary about Flutter), let's look at some of the benefits that put Flutter ahead of its competitors.

DEVELOPER'S PRODUCTIVITY

The 'Hot Reload' (Hot reload) feature helps developers to move faster. This feature allows developers to see the changes reflected at the interface as they write code in real-time. There's no time wasted in the long build-times to be able to check out the changes made.

FRONT-END & BACKEND USING DART

Flutter apps are written using the Dart language. The same programming language can also be used to build the backend for the application. Developing front-end and backend in the same language helps with the cost and developers' familiarity with both sides of the app development. The same developers can own the front-end as well as the backend development that leads to lean and cost-effective development cycles.

DART LANGUAGE

Flutter uses Dart language to write apps. It's a delight to work with Dart language. It is an object-oriented language that is built on the most popular features of other reliable programming languages like Java and similar languages built upon object-oriented concepts. Dart is created, keeping developers in mind, and it is easy to learn enough to start being productive from day one.

HUMAN INTERFACE DESIGN

Flutter team implemented Material Design Specification (Material Foundation) meticulously in Flutter. It makes easy to create slick, smooth, and crisp interfaces for different platforms like Android and iOS. Flutter apps are compiled into Native code for a platform that provides a Native feel on different platforms like Android, iOS, Web, and Desktop.

FIREBASE INTEGRATION

Flutter's integration with Firebase (Firebase Platform) is reliable. Firebase provides the tools and services to support backend infrastructure instantaneously, which is scalable and serverless. It comes with real-time databases, authentication, cloud storage, hosting, cloud-function, machine learning support.

VARIETY IN IDE SUPPORT

Flutter offers multiple integrated development environment (IDE) to develop applications. As a Flutter developer, you have a choice between Android Studio (Google), VS Code (Microsoft), IntelliJ IDEA (Jetbrains), and command line.

OEM WIDGET INDEPENDENCE

Flutter comes with its own suite of widgets that keep your app's interface independent of changes in OEM widgets. OEM widgets come with devices and may look different on different devices. The widgets solve this problem by providing a massive library of widgets for Material Design and iOS (Cupertino) (Cupertino (iOS-style) widgets) platform, which helps in a seamless user experience regardless of which device the app is running on.

OPEN SOURCE/COMMUNITY SUPPORT

Flutter and Dart are free to use open-source, and backed by a very active community of developers worldwide. It's easy to find support, resources, and technical know-how on a topic. Flutter's official documentation is fantastic and accepts contributions from community members.

COST-EFFECTIVENESS

Flutter helps developer in productivity while maintaining low development costs. Flutter is not only used for developing user-facing front-end interfaces, but also solving backend problems. It provides support for code sharing from a single codebase. Flutter apps are built once and deployed on multiple platforms like Android, iOS, web, and desktop. You are not required to have separate teams to support them.

CUSTOM DESIGNS

Flutter is a great choice to build custom designs. It's particularly useful when building designs for the app's branding. Your brand design remains seamless across multiple platforms without worrying about the device. Additionally, it's much faster to develop prototypes that work cross-platform.

CONCLUSION

In this chapter, you were introduced about Flutter and how it compares to other cross-platform alternatives. You learned about the different solutions available for developing applications targeted to various platforms while maximizing the code-reuse. You developed your understanding about Flutter's features and benefits over other cross-platform solutions.

REFERENCES

Adobe. (2020, 11 17). *Adobe PhoneGap*. Retrieved from PhoneGap: https://phonegap.com/
Apple. (2020, 11 16). *Develop iOS Applications*. Retrieved from Apple Developer: https:// developer.apple.com/develop/
Apple. (2020, 11 16). *Swift*. Retrieved from Apple Developer Documentation: https:// developer.apple.com/swift/
Dart Team. (2020, 11 17). *Dart*. Retrieved from Dart Website: https://dart.dev/
Develop Android Applications. (2020, 11 16). Retrieved from Android Developers: https:// developer.android.com/
Flutter Dev. (2020, 12 15). *Hot reload*. Retrieved from flutter.dev: https://flutter.dev/docs/ development/tools/hot-reload
Flutter Team. (2020, 11 17). *Cupertino (iOS-style) widgets*. Retrieved from Flutter Documentation: https://flutter.dev/docs/development/ui/widgets/cupertino

Flutter Team. (2020, 11 17). *Flutter*. Retrieved from GitHub repository for Flutter SDK: https://github.com/flutter

Google. (2020, 11 16). *android studio*. Retrieved from Android Developers: https://developer.android.com/studio

Google. (2020, 11 17). *Material Foundation*. Retrieved from Material Design: https://material.io/design

Google. (2020, 12 15). *Firebase Platform*. Retrieved from Firebase: https://firebase.google.com/

JetBrains & Open-source Contributors. (2020, 11 16). *A modern programming language that makes developers happier*. Retrieved from Kotlin Language: https://kotlinlang.org/

Jetbrains. (2020, 11 16). *IntelliJ IDEA*. Retrieved from Jetbrains: https://www.jetbrains.com/idea/

Leler, W. (2017, 08 20). *What's Revolutionary about Flutter*. Retrieved from hackernoon: https://hackernoon.com/whats-revolutionary-about-flutter-946915b09514

Microsoft. (2020, 11 16). *Visual Studio*. Retrieved from code.visualstudio.com: https://code.visualstudio.com/

Ravn, M. (2018, 08 28). *Flutter Platform Channels*. Retrieved from Medium: https://medium.com/flutter/flutter-platform-channels-ce7f540a104e

Wikipedia. (2020, 11 16). *Java*. Retrieved from Wikipedia: https://en.wikipedia.org/wiki/Java_(programming_language)

Wikipedia. (2020, 11 16). *Objective-C*. Retrieved from Wikipedia: https://en.wikipedia.org/wiki/Objective-C

Wikipedia. (2020, 11 16). *Reactive Programming*. Retrieved from Wikipedia: https://en.wikipedia.org/wiki/Reactive_programming

Wikipedia. (2020, 11 16). *Software development kit*. Retrieved from Wikipedia: https://en.wikipedia.org/wiki/Software_development_kit

3 Setting Up Environment

This chapter gives a quick introduction to setting up a development environment for developing Flutter applications. Flutter applications can be created in macOS, Windows, Linux, and Chrome OS. In this chapter, setting up the environment on macOS is covered. Please refer to the official documentation for setting up the environment on other operating systems here (Install).

The macOS supports developing Flutter apps for all four platforms: Android, iOS, desktop macOS application, and web app in Chrome or Safari. Flutter's desktop support allows it to compile Flutter source code to native macOS or Linux desktop apps. However, a macOS Flutter desktop app can only be assembled on a Mac. A Linux Flutter desktop app can only be compiled/developed on a Linux machine. In this book, we'll be building Flutter app examples for Android, iOS, Web, and macOS desktop applications. You need at least one of the setups to be able to run your first Flutter app.

The Flutter is under rapid development itself, and some of the features like support to web and desktop are not rolled into stable release yet. The support for web is available in 'beta' channel, and desktop support is available in 'dev' channel. You may want to keep your Flutter channel to 'dev' when developing for web and desktop at the same time. The 'master' channel has the bleeding edge improvements, features, and bug fixe. However, this channel is most unstable as well, and hence needs to be avoided in production.

SYSTEM REQUIREMENTS FOR macOS

You need a 64-bit macOS with 2.8 GB disk space only for Flutter installation. Other development tools and Integrated Development Environment (IDE) would require additional disk space. You should be comfortable using command-line tools. Flutter requires the following command-line tools preinstalled in your environment.

1. `bash`: Shell for macOS.
2. `which`: Unix based command to locate executables in the system.
3. `zip` & `unzip`: For compressing and decompressing files and directories.
4. `mkdir` & `rm`: For creating and removing directories and/or files.
5. git (version 2 or greater) (Download for macOS): Tool for version control.
6. curl (command line tool and library): Command line tool for data transfer.

Once you've all these tools installed and ready, it's time to install the Flutter software development kit (SDK) in your environment.

SETTING UP FLUTTER SDK

At the time of this writing, the Flutter SDK's stable version for macOS is 1.22.4 (Flutter Stable 1.22.4). For the latest information about installing Flutter SDK, refer to the Flutter documentation (Get the Flutter SDK). You don't need to worry about installing Dart because Dart SDK is bundled with Flutter SDK.

In this section, we will briefly cover setting up Flutter SDK in the macOS environment. The Flutter is in active development, and a lot of changes are expected. Please refer to the Flutter documentation to setup your Flutter environment for detailed and latest information. I'm describing the installation process at a higher level, which is less likely to change.

DOWNLOAD STABLE FLUTTER SDK

It's always a good idea to install the stable version for Flutter SDK to avoid unexpected behaviors. However, there're many features related to web and desktop variants may not be available in the stable release. The good news is that you're free to change from one version of Flutter to another using Flutter channels.

UPDATE PATH

Don't forget to add Flutter SDK to your environment using `PATH (Update your path). Once you've added the Flutter installation directory to your system's `PATH` variable, you can execute the `flutter` command from the terminal (bash).

FLUTTER DOCTOR

The `flutter doctor` command helps to find out if there are any required dependencies that are not installed yet. If any dependency or tool is missing, it'll let you know about it and tell you how to do it.

FLUTTER CHANNELS

Flutter channels are used to switch from one Flutter version to another. You can choose to switch between different Flutter channels using command `flutter channel` from the command line.

```
```
$ flutter channel
Flutter channels:
* master
  dev
  beta
  stable
```
```

There are four channels at this time that you can choose from. You can switch to the preferred channel using the command `flutter channel <channel _ name>`.

- `master`: You can switch to the master channel using command `flutter channel master`. It contains bleeding-edge changes. You may need to switch to it to run some of the preview features. We may need to switch to this channel when demonstrating the latest and greatest features that haven't been released into stable channels yet.

- `dev`: This channel contains the latest fully-tested build. You can switch to this channel using command `flutter channel dev`.
- `beta`: This channel contains the most stable dev build. It's updated monthly. You can switch to this channel using command `flutter channel beta`.
- `stable`: This channel contains the most stable beta build. It's updated quarterly. You can switch to this channel using command `flutter channel stable`.

After switching to your preferred channel, you need to run the command `flutter upgrade` to bring in updates.

SETTING UP FOR THE ANDROID PLATFORM

ANDROID STUDIO

Next step is to install Android Studio (Install Android Studio) along with the latest Android SDK, Android SDK command-line tools, Android Build-tools.

ANDROID EMULATOR

Create an emulator using Android Virtual Device (AVD) Manager in Android Studio (Android Studio > Tools > AVD Manager). Create a virtual device by clicking on the 'Create Virtual Device' button. Choose hardware, system image (×64), and AVD name for your emulator. Clicking on 'Finish' will create an AVD in the AVD manager. Click on the horizontal triangle icon to start it. Verify it by running command `flutter devices` to see it in the device listing in the terminal.

You can also see device listing using the `flutter devices` command as below:

```
```
$ flutter devices
1 connected devices:

Android SDK built for x86 (mobile) • emulator-5554 •
android-x86 • Android 10 (API 29) (emulator)
```
```

TEST YOUR SETUP

Create a test project using `flutter create testapp`, and run it as below:

```
```
flutter create testapp
cd testapp
flutter run -d android
```
```

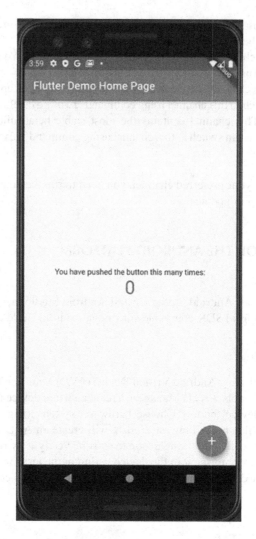

FIGURE 3.1   TestApp running in Android

Refer to Figure 3.1 to see the default app running in the Android emulator.

ANDROID DEVICE

An Android device running Android 4.1 (API level 16) or higher is required to run a Flutter app on an Android device. You would need to connect the device with a USB cable to your computer to access it. Enable USB debugging (Configure on-device developer options) on the Android device. Test the connection using the `flutter devices` command. It should list the device on the console along with other connected devices, if any.

Refer to Flutter Documentation (Android setup) for detailed Android setup instructions.

SETTING UP FOR THE iOS PLATFORM

XCODE

Install the latest stable version of Xcode either from the Mac App Store (Xcode) or the web (Xcode). Follow the prompts for installation. Xcode installation will install command-line tools as well. Refer to Flutter documentation (Install Xcode) for detailed instructions on installing and setting up Xcode.

iOS SIMULATOR

On Mac, open the Simulator using command `open -a simulator`. At the time of this writing, the default simulator is iPhone SE (2nd generation) (iPhone SE (2nd generation)). Refer to documentation (Set up the iOS simulator) for detailed instructions on setting up an iOS simulator. You can also see device listing using the `flutter devices` command as below:

```
```
$ flutter devices
2 connected devices:

Android SDK built for x86 (mobile) • emulator-5554
• android-x86 • Android 10 (API 29) (emulator)
iPhone SE (2nd generation) (mobile) • FCDB9B21-344C-4735-8394-
93F0CF871DB2 • ios   •
com.apple.CoreSimulator.SimRuntime.iOS-13-5 (simulator)
```
```

TEST YOUR SETUP

Create a test project using `flutter create testapp`, and run it as shown below using the simulator identifier. Alternatively, you can choose to run the command `flutter run` to run *testapp* on all connected platforms.

```
```
flutter create testapp
cd testapp
flutter run -d FCDB9B21-344C-4735-8394-93F0CF871DB2
```
```

Refer to Figure 3.2 to see the default app running in the Android emulator.

iOS DEVICE

You can alternatively run the *testapp* on an iOS device. Follow the directions (Deploy to iOS devices) to deploy to iOS devices. You would need to install a third-party

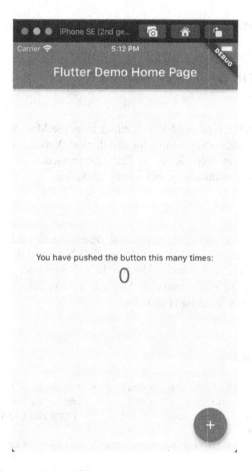

FIGURE 3.2   TestApp running in iOS

dependency manager CocoaPods (CocoaPods) to manage dependencies. You also need an Apple Developer account to run iOS apps on a device. Also, you need to be enrolled in the Apple Developer Program (Apple Developer Program) to release and distribute apps on the App Store. Refer to the setup directions (iOS setup) for detailed iOS setup instructions.

SETTING UP FOR WEB

Flutter's support for the Web is still in the early stages at the time of this writing. Support for the web is available in the beta channel at this time. Switch to beta channel using `flutter channel beta`. Don't forget to run the `flutter upgrade` command and enable web support.

```
```

flutter channel beta
flutter upgrade

FIGURE 3.3   Android studio's 'Run' toolbar showing chrome devices

```
flutter config --enable-web
```

Now that the web is enabled for your environment, you should see 'Chrome (web)' and 'Web Server (web)' listed in your Android Studio's device listing section in the top toolbar.

Refer to Figure 3.3 to see Android studio's 'Run' toolbar showing chrome devices.

You can also see device listing using the `flutter devices` command as below:

```
$ flutter devices
2 connected devices:

Web Server (web)  • web-server  • web-javascript  • Flutter Tools
Chrome (web)      • chrome      • web-javascript  • Google Chrome
84.0.4147.89
```

TEST YOUR SETUP

Create a test project using `flutter create testapp`, and run it as below:

```
flutter create testapp
cd testapp
flutter run -d chrome
```

Refer to Figure 3.4 to see the default app running in the chrome browser.

Refer to Flutter documentation (Building a web application with Flutter) for detailed web setup instructions.

SETTING UP FOR DESKTOP

Flutter's support for the desktop is still in pre-early stages at the time of this writing. Support for the macOS and Linux desktop applications is available in the 'dev' channel at this time. Switch to 'dev' channel using `flutter channel dev`. Don't forget to run the `flutter upgrade` command and enable desktop support.

FIGURE 3.4   TestApp running in chrome

```
```
$ flutter channel dev
$ flutter upgrade
//For macOS
$ flutter config --enable-macos-desktop
//For Linux
$ flutter config --enable-linux-desktop
```
```

Note: You'll see Linux devices only on a Linux machine.

Now that desktop is enabled for your environment, you should see 'macOS (Desktop)' in addition to 'Chrome (web)' and 'Web Server (web)' listed in your Android Studio's device listing section in the top toolbar.

Refer to Figure 3.5 to see Android studio's 'Run' toolbar showing macOS(desktop) and chrome devices.

You can also see device listing using the `flutter devices` command as below:

FIGURE 3.5   Android Studio run toolbar showing macOS desktop and web devices in the device listing

```
```
$ flutter devices
3 connected devices:

macOS (desktop) • macos   • darwin-x64   • Mac OS X
10.15.6 19G73
Web Server (web) • web-server   • web-javascript • Flutter Tools
Chrome (web)   • chrome   • web-javascript • Google Chrome
84.0.4147.105
```
```

TEST YOUR SETUP

Create a test project using `flutter create testapp`, and run it as below:

```
```
flutter create testapp
cd testapp
flutter run -d macos
```
```

Refer to Figure 3.6 to see the default app running in the macOS desktop application.

Refer to the documentation (Flutter Dev, 2020) for detailed desktop setup instructions.

SOURCE CODE ONLINE

This app's source code for all four platforms is available online at GitHub in the 'testapp' project (testapp).

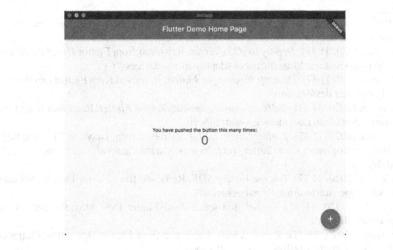

FIGURE 3.6   TestApp running in macOS

SETTING UP EDITOR

You can choose to build apps in Flutter using any editor or just command line (Terminal in Mac) along with Flutter command-line tools. You also have the option to choose the IDE of your choice along with Flutter editor plugins. The Flutter plugin comes with productivity tools like syntax highlighting, easy and intuitive run and debug support from IDE, etc. Few options of editors are Android Studio, IntelliJ, Visual Studio Code, and Emacs. I'm choosing Android Studio to run the code snippets and examples in this book. I chose the Android Studio because it comes with all the tools needed for Android development. You will need Xcode to prepare your iOS app to distribute to App Store covered later in this book.

CONCLUSION

In this chapter, you got pointers for setting up the development environment on MacOS. You got a high-level view of setting up a development environment for building Flutter applications for Android, iOS, web, and desktop (macOS) platforms. You learned to test platform setup for each of the platforms. You got an insight into IDE options available for Flutter development as well.

REFERENCES

Android Developers. (2020, 12 14). *Configure on-device developer options*. Retrieved from developer.android.com: https://developer.android.com/studio/debug/dev-options

Apple. (2020, 11 17). *Xcode*. Retrieved from Apple Apps: https://apps.apple.com/us/app/xcode/id497799835

Apple. (2020, 11 17). *Xcode*. Retrieved from Apple Developer: https://developer.apple.com/xcode/

Apple. (2020, 12 15). *Apple Developer Program*. Retrieved from developer.apple.com: https://developer.apple.com/programs/

Cocoapods.org. (2020, 12 15). *CocoaPods*. Retrieved from cocoapods.org: https://cocoapods.org/

Everything curl. (2020, 11 17). *command line tool and library*. Retrieved from Curl Website: https://curl.se/

Flutter Dev. (2020, 11 17). *Deploy to iOS devices*. Retrieved from Flutter Dev: https://flutter.dev/docs/get-started/install/macos#deploy-to-ios-devices

Flutter Dev. (2020, 11 17). *Desktop support for Flutter*. Retrieved from Flutter Development: https://flutter.dev/desktop

Flutter Team. (2020, 11 17). *Building a web application with Flutter*. Retrieved from Flutter Web: https://flutter.dev/docs/get-started/web

Flutter Team. (2020, 11 17). *Flutter Stable 1.22.4*. Retrieved from Download Flutter: https://storage.googleapis.com/flutter_infra/releases/stable/macos/flutter_macos_1.22.4-stable.zip

Flutter Team. (2020, 11 17). *Get the Flutter SDK*. Retrieved from Flutter Dev: https://flutter.dev/docs/get-started/install/macos#get-sdk

Flutter Team. (2020, 11 17). *Install*. Retrieved from Flutter Dev: https://flutter.dev/docs/get-started/install

Flutter Team. (2020, 11 17). *Install Xcode*. Retrieved from Flutter Dev: https://flutter.dev/docs/get-started/install/macos#install-xcode

Flutter Team. (2020, 11 17). *iOS setup*. Retrieved from Flutter Dev: https://flutter.dev/docs/get-started/install/macos#ios-setup

Flutter Team. (2020, 11 17). *Set up the iOS simulator*. Retrieved from Flutter Dev: https://flutter.dev/docs/get-started/install/macos#set-up-the-ios-simulator

Flutter Team. (2020, 11 17). *Update your path*. Retrieved from Flutter Dev: https://flutter.dev/docs/get-started/install/macos#update-your-path

Git. (2020, 11 17). *Download for macOS*. Retrieved from Install Git: https://git-scm.com/download/mac

Google. (2020, 11 17). *Android setup*. Retrieved from Flutter Dev: https://flutter.dev/docs/get-started/install/macos#android-setup

Google. (2020, 11 17). *Install Android Studio*. Retrieved from Flutter Dev: https://flutter.dev/docs/get-started/install/macos#install-android-studio

Priyanka Tyagi. (2020, 11 17). *testapp*. Retrieved from GitHub: https://github.com/ptyagicodecamp/testapp

Wikipedia. (2020, 11 17). *iPhone SE (2nd generation)*. Retrieved from Wikipedia: https://en.wikipedia.org/wiki/IPhone_SE_(2nd_generation)

4 Flutter Project Structure

In this chapter, you'll develop your understanding of the Flutter project's structure. We'll be creating a Flutter app 'Hello Books' for demonstration. The *'Hello Books'* app will display this greeting in three different languages: Spanish (Hola Libros), Italian (Ciao Libri), and Hindi (हैलो किताबें). The default greeting is displayed in the English language. All other three greetings are stored in a List (List<E> class) data structure. The app has a round button to select the greeting randomly from this list.

Figure 4.1 shows the finished 'Hello Books' app. Clicking on 'Smiley' icon switches the greeting to a different language.

CHOOSING FLUTTER CHANNEL

We'll be using the *'dev'* channel to run the examples in this book since early support for desktop along with support for web, Android, and iOS is available in this channel. After switching to the *'dev'* channel, don't forget to run `flutter upgrade`. You should see the following output on the console:

```
```
Flutter 1.21.0-1.0.pre • channel dev •
https://github.com/flutter/flutter.git
Framework • revision f25bd9c55c (2 weeks ago) • 2020-07-14
20:26:01 -0400
```

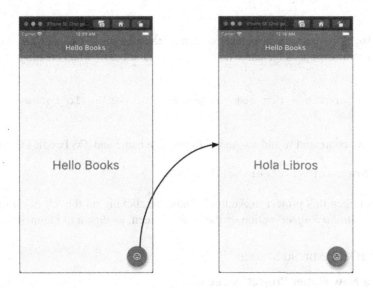

**FIGURE 4.1**    Hello Books App (Finished). Clicking on the 'Smiley' icon changes the greeting
```

```
Engine • revision 99c2b3a245
Tools • Dart 2.9.0 (build 2.9.0-21.0.dev 20bf2fcf56)
```

You can run the `flutter channel` command to see the active channel marked with '*'.

```
$ flutter channel
Flutter channels:
  master
* dev
  beta
  stable
```

CREATING FLUTTER PROJECT

There are two ways to create Flutter projects: using command-line tools from Terminal or using IDE. You can also create the project using the command line and open it from IDE later on.

USING COMMAND LINE (TERMINAL)

Run the following command to create a '*hello_books*' project for the '*Hello Books*' application.

```
$ flutter create hello_books
```

By default, the package structure is `com.example.hello _ books`. You can specify your own package name using the command below when creating a project.

```
flutter create --org com.pragmatic_flutter hello_books
```

The above command would set Android's package name and iOS bundle identifier to

`com.pragmatic_flutter.hello_books`.

You can open this project in Android Studio by clicking on the '*Open an existing Android Studio Project*' option on the launch screen, as shown in Figure 4.2.

USING IDE (ANDROID STUDIO)

'Start a New Flutter Project' Screen
Start a new project by selecting this option on the launch screen as shown in Figure 4.3.

FIGURE 4.2 Android Studio launch screen displaying 'Open an existing Android Studio Project' option

FIGURE 4.3 'Start a New Flutter Project' screen

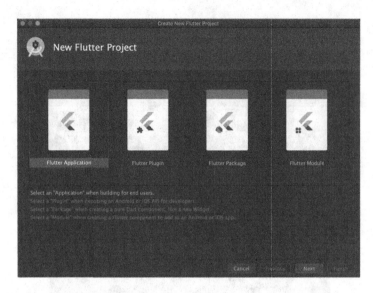

FIGURE 4.4   'New Flutter Project' screen

'New Flutter Project' Screen

This screen gives you the option to create a Flutter project, a Flutter module, package, or plugin. Since we're creating a Flutter application, so choose the 'Flutter Application' option as shown in Figure 4.4.

'New Flutter Application' Screen

Configure options for the new Flutter application on this screen shown in Figure 4.5.

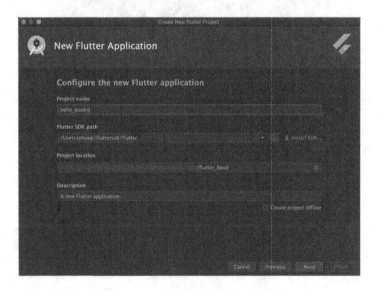

FIGURE 4.5   'New Flutter Application' screen (configuring project)

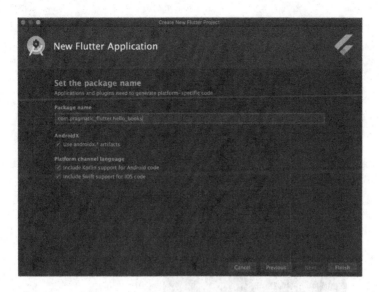

FIGURE 4.6   'New Flutter Application' screen (setting package name)

Next, provide the package name to organize the source code. You also need to select the support for '*andoidx artifacts*' and Kotlin and Swift language support for Android and iOS platforms as displayed in Figure 4.6.

CROSS-PLATFORM FLUTTER PROJECT STRUCTURE

The Flutter project has shared code in the '*lib*' folder. Each platform has its own folder to keep the platform-specific code. There are '*android*', '*iOS*', '*web*', '*linux*', and '*macOS*' folders for each platform Flutter app can be deployed to. The visual for project structure is shown in Figure 4.7.

Let's check out the details of significant project directories and files below:

- `lib/`: This folder contains all Dart files. This is a shared code across all platforms like iOS, Web, Desktop, and embedded devices. In this book, we're focusing on Android, iOS, Web, and Desktop (macOS) only.
- `android/`: This folder contains native Android code, including `AndroidManifest.xml`.
- `ios/`: This folder contains the iOS application's native code in the Swift language.
- `web/`: This folder includes '`index.html`', assets that can be deployed to the public hosting site.
- `linux/`: This folder contains the Linux platform-related code. This folder is generated by enabling the support for desktop (Linux) but can only be compiled on a Linux machine.
- `macos/`: This folder contains native code for the macOS platform.

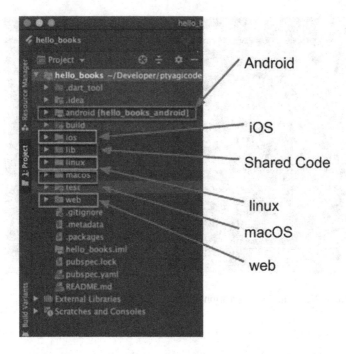

FIGURE 4.7   Flutter project structure in Android Studio

- `test/`: This folder contains all unit testing classes.
- `pubspec.yaml`: This is the dependency management and configuration file for the Flutter application.

RUNNING DEFAULT APP: ANDROID, iOS, WEB, AND DESKTOP

In this section, we will run the '*hello_books*' Flutter application on all four platforms: Android, iOS, web, and desktop (macOS). Let's run the '*hello_books*' project on all four platforms together using one simple command, 'flutter run -d all'.

ANDROID PLATFORM

The Hello Books app running on Android emulator is shown in Figure 4.8.

iOS PLATFORM

The Hello Books app running on iOS simulator is shown in Figure 4.9.

WEB PLATFORM

The Hello Books app running in Chrome browser is shown in Figure 4.10.

FIGURE 4.8   Hello Books running on Android emulator

DESKTOP (macOS) PLATFORM

The Hello Books app running on macOS platform is shown in Figure 4.11.

SOURCE CODE ONLINE

Source code for the '*hello_books*' project is available at GitHub (Chapter04: Flutter Project Structure (hello_books)).

RUNNING CODE SAMPLES

Flutter's documentation is developer-friendly and easy to follow. It comes with a vast library of sample code snippets. It provides the support to create Flutter applications using sample code IDs and run it as a standalone Flutter app.

FIGURE 4.9   Hello Books running on iOS simulator

LISTING SAMPLE CODE IDS

The following command dumps the sample code identifiers/IDs in file *'samples.json'*.

```
flutter create --list-samples=samples.json
```

CONTENTS OF FILE *'samples.json'*

Below is a small snippet of the file. This particular sample code demonstrates the usage of a `BottomAppBar` widget and a docked `FloatingActionButton` in it. You'll learn about the widgets and building interfaces in upcoming chapters. The purpose of this section is to learn to explore Flutter's official sample code on your own.

FIGURE 4.10 Hello Books running on the web platform

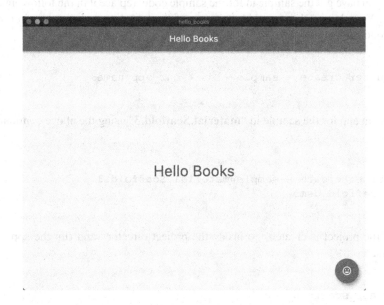

FIGURE 4.11 Hello Books running on macOS platform

```
```
[
{
  "sourcePath": "lib/src/material/scaffold.dart",
  "sourceLine": 1009,
  "id": "material.Scaffold.3",
  "channel": "master",
  "serial": "3",
  "package": "flutter",
  "library": "material",
  "element": "Scaffold",
  "file": "material.Scaffold.3.dart",
  "description": "This example shows a [Scaffold] with an
[AppBar], a [BottomAppBar] and a\n[FloatingActionButton]. The
[body] is a [Text] placed in a [Center] in order\nto center the
text within the [Scaffold]. The [FloatingActionButton]
is\ncentered and docked within the [BottomAppBar] using\
n[FloatingActionButtonLocation.centerDocked]. The
[FloatingActionButton] is\nconnected to a callback that
increments a counter.\n\n![](https://flutter.github.io/assets-
for-api-docs/assets/material/scaffold_bottom_app_bar.png)"
},

]
```
```

CREATING APPS FROM SAMPLE CODE

Once you have got the sample id for the sample code, replace it in the following command. You also need to provide the name of the sample app that you're importing this code snippet into.

```
$ flutter create --sample=<id> <your_app_name>
```

Create an app for the sample id "**material.Scaffold.3**" using the above command as below:

```
$ flutter create --sample=material.Scaffold.3
app_scaffold_demo
```

Once the project is created, go inside the project directory and run the app on all targets.

```
$ cd app_scaffold_demo
$ flutter run -d all
```

FIGURE 4.12 Creating app from sample code ID (docked FloatingActionButton in BottomAppBar widget)

App created using the sample code id 'material.Scaffold.3'. This app shows a docked 'FloatingActionButton' (FloatingActionButton class) in the 'BottomAppBar' (BottomAppBar class) widget as shown in Figure 4.12. We will learn more about the Flutter application's anatomy and widgets in the upcoming chapters.

Source Code Online

Source code for the 'app _ scaffold _ demo' project is available at GitHub (Chapter04: Flutter Project Structure).

USEFUL COMMANDS

Lastly, let's reiterate the essential and useful commands used in Flutter development as below:

- `flutter doctor`: Checks if your machine has all the needed packages and software to build flutter apps.
- `flutter create`: Generates new flutter app.

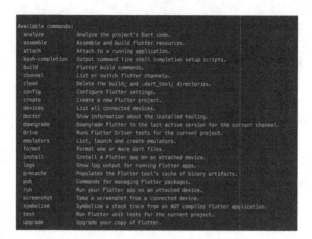

FIGURE 4.13   Available commands for Flutter

- `flutter build`: Builds flutter app.
- `flutter run`: Run flutter app on an attached device. It gives you the option to select a device from the connected devices.
- `flutter help`: You can run `'flutter help'` in the command line to list all other available commands. A screenshot of available commands is shown in Figure 4.13.

CONCLUSION

In this chapter, you learned to create the Flutter project from the command line as well as from the Android Studio IDE. You learned to open a preexisting Flutter project from Android Studio. You learned about the Flutter release channels and when to pick the '*dev*' channel over the '*stable*' or '*beta*' channels. Finally, you also learned how to create a Flutter project from sample code and run it. In the next chapter (Chapter 05: Flutter App Structure), you'll learn about the anatomy and app structure of the 'Hello Books' Flutter application.

REFERENCES

Dart Team. (2020, 11 17). *List<E> class*. Retrieved from Dart Dev: https://api.dart.dev/ stable/2.8.4/dart-core/List-class.html

Flutter Dev. (2020, 11 17). *BottomAppBar class*. Retrieved from Flutter Dev: https://api. flutter.dev/flutter/material/BottomAppBar-class.html

Flutter Dev. (2020, 11 17). *FloatingActionButton class*. Retrieved from Flutter Dev: https:// api.flutter.dev/flutter/material/FloatingActionButton-class.html

Tyagi, P. (2020, 11 17). *Chapter04: Flutter Project Structure*. Retrieved from Pragmatic Flutter GitHub Repo: https://github.com/ptyagicodecamp/app_scaffold_demo

Tyagi, P. (2020, 11 17). *Chapter04: Flutter Project Structure (hello_books)*. Retrieved from Pragmatic Flutter GitHub Repo: https://github.com/ptyagicodecamp/hello_books

Tyagi, P. (2021). Chapter 05: Flutter App Structure. In P. Tyagi, *Pragmatic Flutter: Building Cross-Platform Mobile Apps for Android, iOS, Web & Desktop*. CRC Press.

5 Flutter App Structure

In this chapter, you will develop your understanding of the Flutter application's structure using the *HelloBooksApp* from the previous chapter (Chapter 04: Flutter Project Structure). The app starts displaying a greetings text 'Hello Books' in the center of the device's screen. This app will display this greeting in three different languages: Spanish (Hola Libros), Italian (Ciao Libri), and Hindi (हैलो किताबें), in addition to the English language as the default. All other three greetings are stored in a List (List<E> class) data structure `greetings`.

```
final List<String> greetings = [
  'Hello Books',
  'Hola Libros',
  'Ciao Libri',
  'हैलो किताबें',
];
```

The app has a round floating button in the bottom right corner with a smiley icon to select the next message from this list. Figure 5.1 shows transitioning from English text to the next Spanish text. Each click on the smiley button will fetch the next message in the list. After reaching the last message in the list, it will resume from the first message in the `greetings` list.

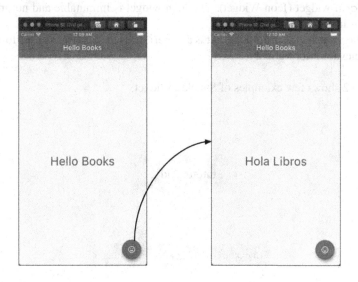

FIGURE 5.1   Finished HelloBooksApp

FLUTTER WIDGETS

STATELESSWIDGET

A StatelessWidget (StatelessWidget class) widget is immutable. Once it's created, it can't be changed. It doesn't preserve its state. A stateless widget is created only once, and it cannot be rebuild at a later time. Stateless widget's 'build()' method is called only once.

Let's use the Container (Container Widget) widget to understand creating a custom StatelessWidget. A container widget is a convenience widget that combines common painting, positioning, and sizing widgets.

The following code snippet demonstrates how a Container widget is created in a stateless manner. This widget is created only once.

```
class MyStatelessWidget extends StatelessWidget {
  @override
  Widget build(BuildContext context) {
    return Container();
  }
}
```

Stateless widgets don't keep track of their state. Once they are created, their value can't be changed. If the value of the stateless widget needs to be changed, then new widgets need to be created with the updated value. Some of the examples of Stateless widgets are below:

- Text widget (Text Widget): It contains the immutable string of letters.
- Icon widget (Icon Widget): The icon widget is immutable and not meant for interaction.
- Card widget (Card Widget): It is a material design card that is used to show relevant information.

Figure 5.2 shows few examples of StatelessWidget.

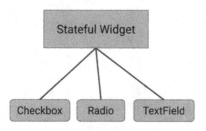

FIGURE 5.2   Examples of StatelessWidget

STATEFULWIDGET

A StatefulWidget (StatefulWidget) widget is mutable. It keeps track of the state. It rebuilds several times over its lifetime. A Stateful widget's `build()` method is called multiple times.

```
class MyStatefulWidget extends StatefulWidget {
  @override
  _MyStatefulWidgetState createState() =>
_MyStatefulWidgetState();
}
class _MyStatefulWidgetState extends State<MyStatefulWidget> {
  @override
  Widget build(BuildContext context) {
    return Container();
  }
}
```

Some of the examples of Stateful widgets are below. They're meant to keep track of their state.

- Checkbox (Checkbox class): Keeps track of its state whether a checkbox is checked or not.
- Radio (Radio<T> class): Needs to remember its state if it's selected or not.
- TextField (TextField class): The TextField widgets enable users to type in the text. That means it needs to keep track of what the user has already typed.

Figure 5.3 shows examples of StatefulWidget.

In the *HelloBooksApp*, we'll use StatefulWidget to implement the 'Smiley' round button to change the greeting display text.

DISPLAY 'HELLO BOOKS' TEXT

We will start building the *HelloBooksApp* by displaying the 'Hello Books' text. The app makes use of Material Design (Material Design). The MaterialApp

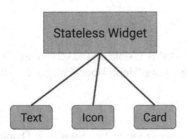

FIGURE 5.3 Examples of StatefulWidget

(MaterialApp class) widget is used to implement material design. The class `HelloBooksApp` extends `StatelessWidget`. A Flutter application is a stateless widget because it's immutable. It can have mutable and immutable widgets as its children. The `HelloBooksApp` widget implements the `build()` method of parent `StatelessWidget`. The `build()` method is responsible for composing the widgets to build the user interface. In this `build()` method, MaterialApp widget is returned. The Text (Text class) widget is used to display the 'Hello Books' text. Let's create a MaterialApp and display 'Hello Books' text in the code below:

```
class HelloBooksApp extends StatelessWidget {
    // This widget is the root of your application.
    @override
    Widget build(BuildContext context) {
      return MaterialApp(
        debugShowCheckedModeBanner: false,
        //Text Widget without SafeArea
        home: Text('Hello Books'),
      );
    }
}
```

The Flutter app during development has a debug banner at the top right corner. You can remove it by setting the flag `debugShowCheckedModeBanner` in MaterialApp to `false`.

ENTRY POINT

We have HelloBooksApp ready to run on devices. We need to specify an entry point where the compiler can start executing the app code. The `main()` method in most of the programming languages is the method from which the compiler starts execution.

```
void main() {
    runApp(HelloBooksApp());
}
```

OUTPUT

The text string 'Hello Books' is visible on the screen, as shown in Figure 5.4. You'll see the Text widget displaying 'Hello Books' aligned to the top of the screen.

SOURCE CODE ONLINE

Source code for this example (Flutter App Structure: Display 'Hello Books' text) is available at GitHub.

FIGURE 5.4 'Hello Books' displayed without SafeArea

ADD CUSHION AROUND THE TEXT

At this point, we have the text 'Hello Books' glued to the top of the screen. The top part of the screen is used for showing system notifications. The Text widget overlapping system notification is not a good design and usability decision. This is where the SafeArea (Google, 2020) widget comes in handy. Wrapping Text widget in SafeArea provides the safety padding to avoid the operating system level notifications.

```
```
class HelloBooksApp extends StatelessWidget {
 @override
 Widget build(BuildContext context) {
   return MaterialApp(
     debugShowCheckedModeBanner: false,
     home: SafeArea(
```
```

```
        child: Text('Hello Books'),
     ),
   );
 }
}
```

OUTPUT

The 'Hello Books' text is padded from the top and provides enough space for the operating system notifications.

Figure 5.5 shows the 'Hello Books' text rendered in a SafeArea widget.

SOURCE CODE ONLINE

Source code for this example (Flutter App Structure: Add cushion around text) is available at GitHub.

FIGURE 5.5   'Hello Books' displayed wrapped in SafeArea widget

CENTER THE TEXT

Now that we have 'Hello Books' rendered with enough padding from the top, we want to display text in the middle of the screen. The `Center` (Center class) widget centers its child. Wrapping the `Text` (Text class) widget in the `Center` widget displays the 'Hello Books' in the screen's center.

CODE

```
```
class HelloBooksApp extends StatelessWidget {
 @override
 Widget build(BuildContext context) {
   return MaterialApp(
     debugShowCheckedModeBanner: false,
     home: SafeArea(
       child: Center(
         child: Text('Hello Books'),
       ),
     ),
   );
 }
}
```
```

OUTPUT

The 'Hello Books' text is centered in the middle of the screen.

Refer to Figure 5.6 to see 'Hello Books' and its parent widget Center wrapped in SafeArea widget.

SOURCE CODE ONLINE

Source code for this example (Flutter App Structure: Center the text) is available at GitHub.

APP ANATOMY #1

The *HelloBooksApp* is made of `MaterialApp`, `SafeArea`, `Center`, and `Text` widgets. The `MaterialApp` widget is at the root level. `SafeArea` widget is added as its child. `SafeArea` widget's child is `Center`, which wraps a `Text` widget in it. Figure 5.7 shows this relationship visually.

APP ANATOMY #2

Let's improvise the app to look more like the finished app. We'll be adding new widgets like `Scaffold` (Scaffold class), `AppBar` (AppBar class), and `FloatingActionButton` (FloatingActionButton) next.

Figure 5.8 demonstrates the final version of Hello Books app.

FIGURE 5.6 'Hello Books' wrapped in SafeArea and Center widget. Center widget aligns its child in the center of the screen

THE Scaffold WIDGET

The MaterialApp widget is the starting point of the app. It's used to inform Flutter that the app is going to use material components (Components). Material components are interactive building blocks to create a user interface. The Scaffold widget is used as the child to MaterialApp widget. It implements the material design basic visual layout structure and provides basic functionalities like app bar, floating action button, etc., for the app.

```
```
class HelloBooksApp extends StatelessWidget {
  @override
  Widget build(BuildContext context) {
    return MaterialApp(
      home: Scaffold(),
    );
  }
}
```
```

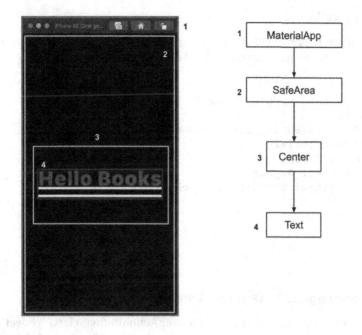

FIGURE 5.7 Anatomy#1 – Hello Books Flutter App

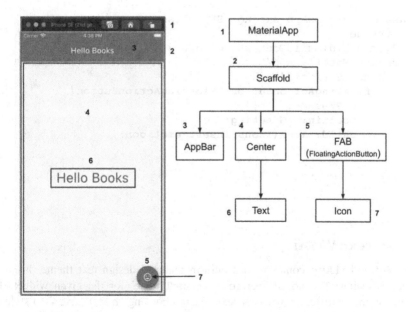

FIGURE 5.8 Anatomy#2 – Hello Books Flutter App

THE AppBar WIDGET

The AppBar (AppBar class) widget implements the material design app bar. It assigns a title to the page using the `title` property.

```
class HelloBooksApp extends StatelessWidget {
@override
Widget build(BuildContext context) {
  return MaterialApp(
    home: Scaffold(
      appBar: AppBar(
        title: Text('Hello Books'),
      ),
    ),
  );
}
```

THE FloatingActionButton WIDGET

The FloatingActionButton (FloatingActionButton class) widget implements the material design floating action button. The Scaffold widget's `floatingActionButton` property assigns FloatingActionButton to the app. The smiley icon is implemented using the Icon widget as a child to FloatingActionButton.

```
class HelloBooksApp extends StatelessWidget {
@override
Widget build(BuildContext context) {
  return MaterialApp(
    home: Scaffold(
      floatingActionButton: FloatingActionButton(
        onPressed: () {},
        tooltip: 'Greeting',
        child: Icon(Icons.insert_emoticon),
      ),
    ),
  );
}
```

STYLING Text WIDGET

The MaterialApp comes with a built-in material design text theme. It can be accessed using `Theme.of(context).textTheme` for the given widget. The theme supports different sizes for text. We're choosing `headline4` to style the text. Themes are covered in detail in a later chapter (Chapter 10: Flutter Themes).

```
~~~
class HelloBooksApp extends StatelessWidget {
@override
Widget build(BuildContext context) {
  return MaterialApp(
    home: Scaffold(
      body: Center(
        child: Text(
          'Hello Books',
          style: Theme.of(context).textTheme.headline4,
        ),
      ),
    ),
  );
}
~~~
```

COMPLETE CODE

The *HelloBooksApp* code is ready to run. The full implementation is shown in the code snippet below:

```
~~~
class HelloBooksApp extends StatelessWidget {
@override
Widget build(BuildContext context) {
  return MaterialApp(
    debugShowCheckedModeBanner: false,
    home: Scaffold(
      appBar: AppBar(
        title: Text('Hello Books'),
      ),
      body: Center(
        child: Text(
          'Hello Books',
          style: Theme.of(context).textTheme.headline4,
        ),
      ),
      floatingActionButton: FloatingActionButton(
        onPressed: () {},
        tooltip: 'Greeting',
        child: Icon(Icons.insert_emoticon),
      ),
    ),
  );
}
}
~~~
```

SOURCE CODE ONLINE

Source code for this example (Flutter App Structure: App Anatomy#2) is available
at GitHub.

MANAGING STATE WITH StatefulWidget

We want to change the greetings text by clicking the smiley floating action button.
First, we need to store all the greetings in a list. Second, we need to enable the floating action button to pick the next available string from the list.

We need StatefulWidget to keep track of the current state of the selected
greeting. Let's go ahead and create a StatefulWidget, say MyHomePage. This
widget is assigned as `home` to MaterialApp.

```
class HelloBooksApp extends StatelessWidget {
  @override
  Widget build(BuildContext context) {
    return MaterialApp(
      ...
      home: MyHomePage(title: 'Hello Books'),
    );
  }
}
```

STATEFULWIDGET: MyHomePage

The Stateful widgets are useful when a part of the screen needs to be updated
with new information. In our app, we want to update the greeting text while rest of
the screen remains unchanged. The MyHomePage extends StatefulWidget and
accepts a title parameter. The widget has a mutable state and represented using
class ` _ MyHomePageState`.

```
class MyHomePage extends StatefulWidget {
  MyHomePage({Key key, this.title}) : super(key: key);
  final String title;
  @override
  _MyHomePageState createState() => _MyHomePageState();
}
class _MyHomePageState extends State<MyHomePage> {}
```

STATE WIDGET: _ MyHomePageState

The widgets are rebuilt whenever the state change is requested from the `set-
State` (setState method) method. Let's see how it's done in *HelloBooksApp*. The

FloatingActionButton is pressed by the user, which updates the currently selected greeting to display on the screen.

```
~~~
class _MyHomePageState extends State<MyHomePage> {
  //Spanish (Hola Libros), Italian (Ciao Libri), and Hindi
  (हैलो किताबें)
  final List<String> greetings = [
    'Hello Books',
    'Hola Libros',
    'Ciao Libri',
    'हैलो किताबें',
  ];

  int index = 0;
  String current;
  void _updateGreeting() {
    setState(() {
      current = greetings[index];
      index = index == (greetings.length - 1) ? 0 : index + 1;
    });
  }

  @override
  Widget build(BuildContext context) {
    return Scaffold(
      appBar: AppBar(
        title: Text(widget.title),
      ),
      body: Center(
        child: Text(
          greetings[index],
          style: Theme.of(context).textTheme.headline4,
        ),
      ),
      floatingActionButton: FloatingActionButton(
        onPressed: _updateGreeting,
        tooltip: 'Greeting',
        child: Icon(Icons.insert_emoticon),
      ),
    );
  }
}
~~~
```

The `greetings` list contains greetings in four languages. The variable `index` keeps the index of the currently selected item in the `greetings` list. On pressing smiley floating action button or FAB calls `_updateGreeting` method. The `_updateGreeting` method updates currently selected greeting text and updates `index` by one. The `index` is reset to zero when it reaches the end of the list.

COMPLETE CODE

The full implementation, including updating the text's language, is in the code below:

```
```

```
//Entry point to the app
void main() {
 runApp(HelloBooksApp());
}

class HelloBooksApp extends StatelessWidget {
 // This widget is the root of your application.
 @override
 Widget build(BuildContext context) {
   return MaterialApp(
     debugShowCheckedModeBanner: false,
     home: MyHomePage(title: 'Hello Books'),
   );
 }
}


class MyHomePage extends StatefulWidget {
 MyHomePage({Key key, this.title}) : super(key: key);
 final String title;
 @override
 _MyHomePageState createState() => _MyHomePageState();
}
class _MyHomePageState extends State<MyHomePage> {
 //Spanish (Hola Libros), Italian (Ciao Libri), and Hindi
 (हैलो किताबे)
 final List<String> greetings = [
   'Hello Books',
   'Hola Libros',
   'Ciao Libri',
   'हैलो किताबे',
 ];

 int index = 0;
 String current;
 void _updateGreeting() {
   setState(() {
     current = greetings[index];
     index = index == (greetings.length - 1)?  0 : index + 1;
   });
 }

 @override
 Widget build(BuildContext context) {
   return Scaffold(
     appBar: AppBar(
       title: Text(widget.title),
```

```
    ),
    body: Center(
      child: Text(
        greetings[index],
        style: Theme.of(context).textTheme.headline4,
      ),
    ),
    floatingActionButton: FloatingActionButton(
      onPressed: _updateGreeting,
      tooltip: 'Greeting',
      child: Icon(Icons.insert_emoticon),
    ),
  );
}
}
```

SOURCE CODE ONLINE

Source code for this example (Flutter App Structure: HelloBooksApp) is available at GitHub.

CONCLUSION

In this chapter, you learned about the Flutter application's structure and anatomy. You learned about the basic building blocks to build *HelloBooksApp*. We started building the app by displaying the text on the screen. First you looked into the bare minimum Flutter app's anatomy. Then the app is improvised using material components. The final *HelloBooksApp* included a FAB to pick a different greeting text from the list. You are also introduced to the basic Flutter widgets like MaterialApp, Container, AppBar, Scaffold, FloatingActionButton, Text, StatelessWidget, and StatefulWidget.

REFERENCES

Dart Dev. (2020, 11 17). *List<E> class*. Retrieved from Dart Dev: https://api.dart.dev/stable/2.8.4/dart-core/List-class.html

Flutter Dev. (2020, 11 17). *AppBar class*. Retrieved from Flutter Dev: https://api.flutter.dev/flutter/material/AppBar-class.html

Flutter Dev. (2020, 11 17). *Card Widget*. Retrieved from Flutter Dev: https://api.flutter.dev/flutter/material/Card-class.html

Flutter Dev. (2020, 11 17). *Center class*. Retrieved from Flutter Dev: https://api.flutter.dev/flutter/widgets/Center-class.html

Flutter Dev. (2020, 11 17). *Components*. Retrieved from Flutter Dev: https://material.io/components

Flutter Dev. (2020, 11 17). *Container Widget*. Retrieved from Flutter Dev: https://api.flutter.dev/flutter/widgets/Container-class.html

Flutter Dev. (2020, 11 17). *FloatingActionButton*. Retrieved from Flutter Dev: https://api.flutter.dev/flutter/material/FloatingActionButton-class.html

Flutter Dev. (2020, 11 17). *FloatingActionButton class*. Retrieved from Flutter Dev: https://api.flutter.dev/flutter/material/FloatingActionButton-class.html

Flutter Dev. (2020, 11 17). *Icon Widget*. Retrieved from Flutter Dev: https://api.flutter.dev/flutter/widgets/Icon-class.html

Flutter Dev. (2020, 11 17). *MaterialApp class*. Retrieved from Flutter Dev: https://api.flutter.dev/flutter/material/MaterialApp-class.html

Flutter Dev. (2020, 11 17). *Scaffold class*. Retrieved from Flutter Dev: https://api.flutter.dev/flutter/material/Scaffold-class.html

Flutter Dev. (2020, 11 17). *setState method*. Retrieved from Flutter Dev: https://api.flutter.dev/flutter/widgets/State/setState.html

Flutter Dev. (2020, 11 17). *StatefulWidget*. Retrieved from Flutter Dev: https://api.flutter.dev/flutter/widgets/StatefulWidget-class.html

Flutter Dev. (2020, 11 17). *StatelessWidget class*. Retrieved from Flutter Dev: https://api.flutter.dev/flutter/widgets/StatelessWidget-class.html

Flutter Dev. (2020, 11 17). *Text class*. Retrieved from Flutter Dev: https://api.flutter.dev/flutter/widgets/Text-class.html

Flutter Dev. (2020, 11 17). *Text Widget*. Retrieved from Flutter Dev: https://api.flutter.dev/flutter/widgets/Text-class.html

Google. (2020, 11 17). *Checkbox class*. Retrieved from Flutter Dev: https://api.flutter.dev/flutter/material/Checkbox-class.html

Google. (2020, 11 17). *Material Design*. Retrieved from material.io: https://material.io/design

Google. (2020, 11 17). *Radio<T> class*. Retrieved from Flutter Dev: https://api.flutter.dev/flutter/material/Radio-class.html

Google. (2020, 11 17). *SafeArea class*. Retrieved from Flutter Dev: https://api.flutter.dev/flutter/widgets/SafeArea-class.html

Google. (2020, 11 17). *Text class*. Retrieved from Flutter Dev: https://api.flutter.dev/flutter/widgets/Text-class.html

Google. (2020, 11 17). *TextField class*. Retrieved from Flutter Dev: https://api.flutter.dev/flutter/material/TextField-class.html

Tyagi, P. (2020, 11 17). *Flutter App Structure: Add cushion around text*. Retrieved from Chapter05: Pragmatic Flutter GitHub Repo: https://github.com/ptyagicodecamp/pragmatic_flutter/blob/master/lib/chapter05/main_05_1.dart

Tyagi, P. (2020, 11 17). *Flutter App Structure: App Anatomy#2*. Retrieved from Chapter05: Pragmatic Flutter GitHub Repo: https://github.com/ptyagicodecamp/pragmatic_flutter/blob/master/lib/chapter05/main_05_3.dart

Tyagi, P. (2020, 11 17). *Flutter App Structure: Center the text*. Retrieved from Chapter05: Pragmatic Flutter GitHub Repo: https://github.com/ptyagicodecamp/pragmatic_flutter/blob/master/lib/chapter05/main_05_2.dart

Tyagi, P. (2020, 11 17). *Flutter App Structure: Display 'Hello Books' text*. Retrieved from Chapter05: Pragmatic Flutter GitHub Repo: https://github.com/ptyagicodecamp/pragmatic_flutter/blob/master/lib/chapter05/main_05_0.dart

Tyagi, P. (2020, 11 17). *Flutter App Structure: HelloBooksApp*. Retrieved from Chapter05: Pragmatic Flutter GitHub Repo: https://github.com/ptyagicodecamp/pragmatic_flutter/blob/master/lib/chapter05/main_05_4.dart

Tyagi, P. (2021). Chapter 04: Flutter Project Structure. In P. Tyagi, *Pragmatic Flutter: Building Cross-Platform Mobile Apps for Android, iOS, Web & Desktop*. CRC Press.

Tyagi, P. (2021). Chapter 10: Flutter Themes. In P. Tyagi, *Pragmatic Flutter: Building Cross-Platform Mobile Apps for Android, iOS, Web & Desktop*. CRC Press.

6 Flutter Widgets

In the previous chapter (Chapter 05: Flutter App Structure), you're briefly introduced to a few basic widgets to create a user interface to display 'Hello Books' text in the middle of the screen. You were introduced to the `MaterialApp`, `Scaffold`, `AppBar`, `Center`, `FAB`, `Text`, and `Icon` widgets. There are many more widgets that are used in real-world apps. In this chapter, we will touch base with more widgets to create more ambitious interfaces.

Image WIDGET

In this section, you'll learn to use the `Image` (Image class) widget to display an image in the Flutter application from the local assets folder as well as over the Internet.

LOCAL IMAGE

In the following example, you will learn to display images from the local folder *assets* at the root level of the project. Create a folder *assets* at the project's root level if you don't have one already. Add an image file that you want to display in the Flutter application's screen. I have file '*flutter_icon.png*' available in the '*assets*' folder. In order to add assets to your application, you need to add an `assets` section under the `flutter` section in the *pubspec.yaml* file.

```
```
flutter:
 assets:
    - assets/flutter_icon.png
```
```

Let's create a method `loadLocalImage()` to return the `Image` widget. This widget is added as the child to the `Container` widget, as shown in the code below:

```
```
Widget build(BuildContext context) {
 return Scaffold(
    appBar: AppBar(
      title: Text("Image Widget"),
    ),
    body: Center(
      child: Container(
        width: 300,
        height: 300,
        padding: const EdgeInsets.all(20.0),
        child: loadLocalImage(),
```
```

```
        ),
      ),
    );
}

Widget loadLocalImage() {
    return Image.asset("assets/flutter_icon.png");
}
```

Figure 6.1 shows the `Image` widget loading image from the local *assets* folder.

INTERNET (REMOTE) IMAGE

The image can also be displayed using the URL. The `loadInternetImage()` fetches and loads images in the `Image` widget using `Image.network()` constructor. This widget is added as the child to the `Container` widget.

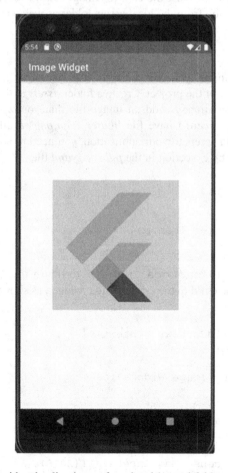

FIGURE 6.1   Image widget loading image from local 'assets' folder

```
` ` `
Widget build(BuildContext context) {
  return Scaffold(
    appBar: AppBar(
      title: Text("Image Widget"),
    ),
    body: Center(
      child: Container(
        width: 300,
        height: 300,
        padding: const EdgeInsets.all(20.0),
        child: loadInternetImage(),
      ),
    ),
  );
}

Widget loadInternetImage() {
    return Image.network( "https://github.com/ptyagicodecamp/
flutter_cookbook2/raw/master/assets/flutter_icon.png");
}
` ` `
```

Figure 6.2 shows the Image widget loading image from the URL.

SOURCE CODE ONLINE

The source code for this example (Flutter Widgets: Image) is available at GitHub.

ToggleButtons WIDGET

In the previous section, we learned to load an image from local assets and from the URL in the Image widget. In this section, we will use ToggleButtons (ToggleButtons class) widget to add two toggle buttons horizontally. The first button is to turn airplane mode off, and another is to turn airplane mode on.

AIRPLANE MODE OFF

When airplane mode is off, applications can access the Internet. The toggle button **airplane_mode_off** loads an image from an Internet URL.
    Refer to Figure 6.3 to see image loading from url.

AIRPLANE MODE ON

When airplane mode is on, the application can't access the Internet and can only access the local assets. The toggle button **airplane_mode_on** loads an image from the local *assets* folder.
    Refer to Figure 6.4 to see image loading locally from 'assets' folder.

FIGURE 6.2   Image widget loading image from the URL

USING `ToggleButtons` WIDGET

The two toggle buttons are added to the layout as below:

```
```
ToggleButtons(
 children: [
   Icon(Icons.airplanemode_off),
   Icon(Icons.airplanemode_on),
 ],
 isSelected: [!isLocal, isLocal],
 onPressed: (int index) {
   setState(() {
     isLocal = !isLocal;
   });
 },
),
```
```

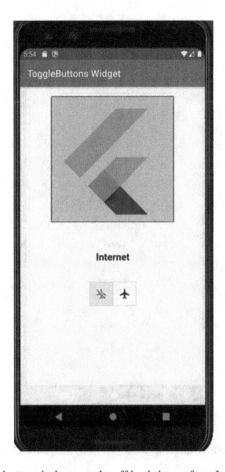

FIGURE 6.3 Toggle button airplane_mode_off loads image from Internet URL

The `isLocal` is a variable of type `bool` that toggles itself every time one of the toggle buttons is selected.

```
```
bool isLocal = true;
class _MyImageWidgetState extends State<MyImageWidget> {
...
Container(
 width: 300,
 height: 300,
 padding: const EdgeInsets.all(20.0),
 child: isLocal == true
     ? Image.asset("assets/flutter_icon.png")
     : Image.network(
         "https://github.com/ptyagicodecamp/flutter_cookbook2/
raw/master/assets/flutter_icon.png"),
),
```
```

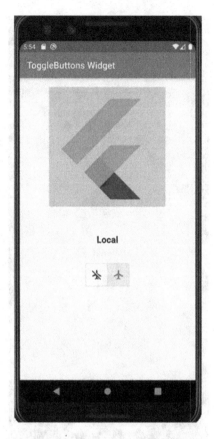

FIGURE 6.4   Toggle button airplane_mode_on loads image from local 'assets' folder

```
...
}
```

When `isLocal` is `true`, the image is loaded from the local *assets* folder, otherwise, it's fetched over the network using the URL.

SOURCE CODE ONLINE

The source code for this example (Flutter Widgets: ToggleButtons) is available at GitHub.

TextField WIDGET

The `TextField` (TextField class) widget lets the user enter the text using hardware or an on-screen keyboard. The `TextEditingController` (TextEditingController class) manages the widget by allowing users enter text, submitting, and finally clearing the text. We will be using a `StatefulWidget` for this example. The

`_MyTextFieldWidgetState` holds the state of this widget. It has a reference to the `TextEditingController` as `_controller`. The `userText` variable contains the text entered in the `TextField` widget.

```
class MyTextFieldWidget extends StatefulWidget {
    MyTextFieldWidget({Key key}) : super(key: key);
    @override
    _MyTextFieldWidgetState createState() =>
    _MyTextFieldWidgetState();
}
class _MyTextFieldWidgetState extends State<MyTextFieldWidget> {
    TextEditingController _controller;
    String userText = "";
    ...
}
```

This controller needs to be initialized inside the `initState()` method. Remember to remove the controller in the `dispose()` method to avoid memory leaks.

```
class _MyTextFieldWidgetState extends State<MyTextFieldWidget> {
    ...

    void initState() {
      super.initState();
      _controller = TextEditingController();
    }
    void dispose() {
      _controller.dispose();
      super.dispose();
    }
    ...
}
```

Next, let's add the `TextField` widget as one of the children to the `Column` layout widget. The `TextField` widget's `autofocus` property is set to `true` to prompt users to enter text. The `TextEditingController` reference `_controller` is assigned to the `controller` property. The `onSubmitted` property tells the widget what to do with the entered text. In this example, the entered text `value` is assigned to the `userText` variable. The `TextField` widget is cleared using `_controller.clear()`.

```
class _MyTextFieldWidgetState extends State<MyTextFieldWidget> {
    TextEditingController _controller;
    ...
```

```
Widget build(BuildContext context) {
  return Scaffold(
    ...,
    body: Padding(
      ...,
      child: Column(
        children: [
          TextField(
            autofocus: true,
            controller: _controller,
            onSubmitted: (String value) async {
              setState(() {
                userText = value;
                _controller.clear();
              });
            },
          ),
          ...,
        ],
      ),
    ),
  );
}
}
```

Add a Text widget below the TextField to display the text entered by the user once they click on the 'Done' or 'Check' button/mark depending on the pop-up software/hardware keyboard.

```
class _MyTextFieldWidgetState extends State<MyTextFieldWidget> {
  ...
  Widget build(BuildContext context) {
    return Scaffold(
      ...,
      body: Padding(
        ...,
        child: Column(
          children: [
            ...,
            Text("User entered: $userText"),
          ],
        ),
      ),
    );
  }
}
```

Figure 6.5 shows the user enters text in the TextField widget.

FIGURE 6.5 User entered the text input

Once the user hits the done button or checkmark on the keyboards, the entered text is displayed in the `Text` widget under it. The `TextField` is cleared and ready for the user to input new text as shown in Figure 6.6.

SOURCE CODE ONLINE

The source code for this example (Flutter Widgets: TextField) is available at GitHub.

`FutureBuilder` ASYNC WIDGET

Sometimes applications don't receive the data required to build the interface all at once. It can happen when data is being retrieved over the network from a server remotely. In such cases, information is received asynchronously using Dart's `Future` (Future<T> class) or `Stream` (Asynchronous programming: streams) classes. The

FIGURE 6.6   The entered text is displayed in `Text` widget. The `TextField` is reset

FutureBuilder (FutureBuilder<T> class) is a widget that builds itself based on the snapshot returned from the Future class.

In this example, we will mock two types of Future objects. The first sample Future object is `_futureData`. It'll return an integer '3' after a delay of three seconds.

Future Data Object

```
` ` `
Future<int> _futureData = Future<int>.
delayed(Duration(seconds: 3), () => 3);
` ` `
```

Future Error Object

The second Future object, `_futureError`, returns an error with the message 'Sample error'.

```
` ` `
Future<int> _futureError =
   Future<int>.delayed(Duration(seconds: 3), () => throw
("Sample error"));

` ` `
```

FutureBuilder WIDGET

The FutureBuilder (FutureBuilder<T> class) widget builds widgets for the interface based on the snapshot received from the Future object. The Future object ` _ futureData` is assigned to its `future` property. The `builder` property is given to the constructor with `AsyncSnapshot<int> snapshot`. It uses `snapshot` to build the `futureChild` widget. When the `snapshot` has data, it displays the data received in a Text widget. If an error is received, then an error message is displayed. When the app is still waiting for data to arrive, a circular progress indicator is shown. The `futureChild` is added as a child to the Center widget.

```
` ` `
FutureBuilder<int> (
      future: _futureData,
      builder: (BuildContext context, AsyncSnapshot<int>
      snapshot) {
        Widget futureChild;
        if (snapshot.hasData) {
          //success
          futureChild = Text("Number received is
$ {snapshot.data}");
        } else if (snapshot.hasError) {
          //show error message
          futureChild = Text("Error occurred fetching data
[${snapshot.error}]");
        } else {
          //waiting for data to arrive
          futureChild = CircularProgressIndicator();
        }


        return Center(
          child: futureChild,
        );
      },
),
` ` `
```

The circular progress bar `CircularProgressIndicator()` is presented to the user when waiting for data to arrive as shown in Figure 6.7.

Figure 6.8 displays the data returned asynchronously from the Future object.

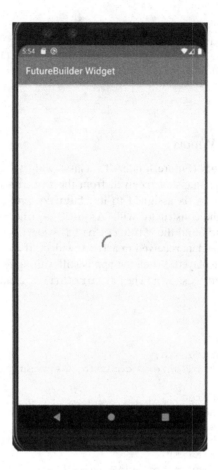

FIGURE 6.7   Circular progress bar while waiting for data to arrive

The source code for this example (Flutter Widgets: FutureBuilder Async Widget) is available at GitHub.

Placeholder WIDGET

In the previous widget `FutureBuilder`, a circular progress bar is displayed on the screen when the app is waiting for data to arrive. When this asynchronous data contains information about an image intended to be displayed, it makes sense to show a placeholder for this image. The `Placeholder` (Placeholder class) widget does precisely that. It draws a box that indicates that a new widget will be added at some point in the future.

Let's create a mock `Future` object that returns the image information delivered asynchronously. The ` _ futureData` is a `Future` object that fetches the name of the asset image to be displayed with a delay of three seconds.

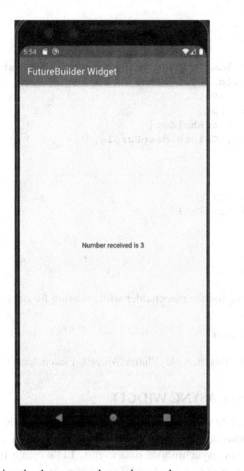

FIGURE 6.8 Displaying the data returned asynchronously

```
Future<String> _futureData = Future<String>.delayed(
   Duration(seconds: 3), () => 'assets/flutter_icon.png');
```

The FutureBuilder widget displays the image using the Image widget when asynchronous data is available, otherwise, it shows a Placeholder widget inside a Container widget.

```
FutureBuilder<String>(
 future: _futureData,
 builder: (BuildContext context, AsyncSnapshot<String> snapshot)
{
   Widget futureChild;
   if (snapshot.hasData) {
```

```
    //success
    futureChild = Image.asset(snapshot.data);
  } else {
    //Placeholder widget while waiting for data to arrive
    futureChild = Container(
      height: 200,
      width: 200,
      child: Placeholder(
        color: Colors.deepPurple,
      ),
    );
  }
  return Center(
    child: futureChild,
  );
},
),
```

Figure 6.9 is displaying the placeholder while waiting for data.

SOURCE CODE ONLINE

The source code for this example (Flutter Widgets: Placeholder) is available at GitHub.

StreamBuilder ASYNC WIDGET

The StreamBuilder (StreamBuilder class) is a widget that builds itself based on the asynchronous data events received from a Stream. A Stream (Stream<T> class) is a source of asynchronous data events. Let's create two mock streams to understand the usage of StreamBuilder widget.

Stream DATA OBJECT

The `_ streamData` is a stream of results/responses that send an integer '3' after a delay of three seconds. It also introduces a delay of additional three seconds after yielding the event.

```
Stream<int> _streamData = (() async* {
  await Future<void>.delayed(Duration(seconds: 3));
  yield 3;
  await Future<void>.delayed(Duration(seconds: 3));
})();
```

Stream ERROR OBJECT

The `_ streamError` yields an error after a delay of three seconds.

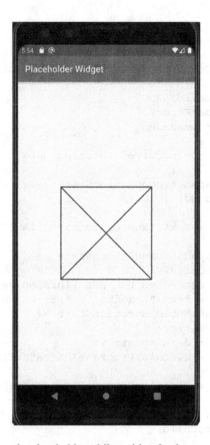

FIGURE 6.9 Displaying the placeholder while waiting for data

```
Stream<int> _streamError = (() async* {
  await Future<void>.delayed(Duration(seconds: 3));
  yield throw ("Error in calculating number");
})();
```

StreamBuilder Widget

The `StreamBuilder` widget reads `_streamData` using its `stream` property. The `builder` takes the asynchronous snapshot events received from the stream and creates a child widget, `futureChild`. If an error is received, the error message is displayed on the screen in a `Text` widget. The `snapshot.connectionState` provides the state of the connection. When the connection state is `ConnectionState.done`, the data received is displayed in the `Text` widget. When the connection state is active, it says "Loading...." on the screen. In any other state, a circular progress indicator widget is shown on the screen.

```
` ` `
StreamBuilder<int>(
 stream: _streamData,
 builder: (BuildContext context, AsyncSnapshot<int> snapshot)
 {
   Widget futureChild;
   if (snapshot.hasError) {
     //show error message
     futureChild =
         Text("Error occurred fetching data
[${snapshot.error}]");
   } else if (snapshot.connectionState ==
ConnectionState.done) {
     //success
     futureChild = Text("Number received is ${snapshot.
data}");
   } else if (snapshot.connectionState ==
ConnectionState.active) {
     //stream is connected but not finished yet.
     futureChild = Text("Loading....");
   } else if (snapshot.connectionState ==
ConnectionState.waiting) {
     //waiting for data to arrive
     futureChild = CircularProgressIndicator();
   } else {
     futureChild = CircularProgressIndicator();
   }
   return Center(
     child: futureChild,
   );
 },
),
` ` `
```

The circular progress bar is displayed while waiting for events in Figure 6.10.
Figure 6.11 shows the stream events' data printed on the screen.

SOURCE CODE ONLINE

The source code for this example (Flutter Widgets: StreamBuilder Async Widget) is available at GitHub.

AlertDialog WIDGET

The AlertDialog (AlertDialog class) is a material design alert dialog. An alert dialog informs the user about situations that require acknowledgment. An alert dialog has an optional title, optional content, and an optional list of actions. The title is displayed above the content, and the actions are shown below the content. This example will demonstrate the alert dialog widget for Android and iOS platforms.

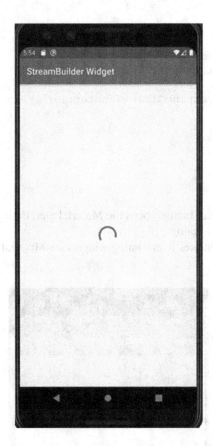

FIGURE 6.10   Circular progress bar while waiting for events

We will add two buttons using `ElevatedButton` widget to show alert dialog for each platform.

ElevatedButton

First, let's add two buttons to open Material and Cupertino (iOS) style alert dialogs. The `ElevatedButton` (ElevatedButton class) widget is used to add buttons to show two variations of `AlertDialog`.

```
```
Center(
 child: Row(
   mainAxisAlignment: MainAxisAlignment.spaceAround,
   children: [
     ElevatedButton(
       child: Text("Material"),
       onPressed: () {
        _showMaterialDialog(context);
```
```

```
        },
      ),
    ElevatedButton(
      child: Text("Cupertino"),
      onPressed: () {
        _showCupertinoDialog(context);
      },
    ),
  ],
  ),
),
~ ~ ~
```

Clicking on the *Material* button opens the Material alert dialog, and *Cupertino* will open an iOS-style alert dialog.

Figure 6.12 demonstrates AlertDialog widgets for Material and Cupertino (iOS) styles.

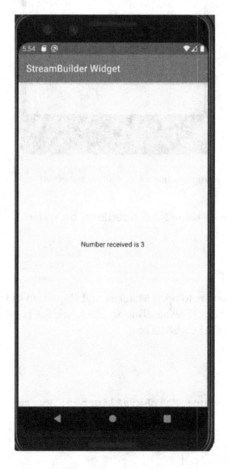

FIGURE 6.11   Displaying stream events

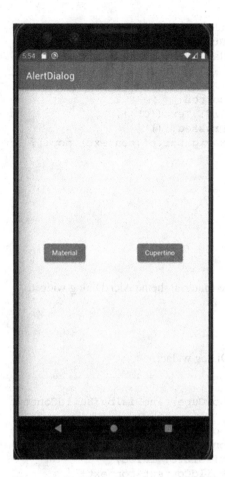

FIGURE 6.12   'ElevatedButton' widgets showing two variations of AlertDialog

MATERIAL STYLE

Material Component `AlertDialog` widget is created as shown in the code snippet below:

```
```
Future<void> _showMaterialDialog(BuildContext context) async
{
 return showDialog<void>(
    context: context,
    barrierDismissible: false,
    builder: (BuildContext context) {
      return AlertDialog(
        title: Text("Material"),
        content: Text("I'm Material AlertDialog Widget."),
        actions: <Widget>[
          TextButton(
```
```

```
            child: Text('Cancel'),
            onPressed: () {
              Navigator.of(context).pop();
            },
          ),
          TextButton(
            child: Text('OK'),
            onPressed: () {
              Navigator.of(context).pop();
            },
          ),
        ],
      );
    });
}
```

```
```

Figure 6.13 shows the Material theme AlertDialog widget.

CUPERTINO STYLE

The iOS style AlertDialog widget.

```
```

```
Future<void> _showCupertinoDialog(BuildContext context) async
{
  return showDialog<void>(
      context: context,
      barrierDismissible: false,
      builder: (BuildContext context) {
        return CupertinoAlertDialog(
          title: Text("Cupertino"),
          content: Text("I'm Cupertino (iOS) AlertDialog
Widget."),
          actions: <Widget>[
            TextButton(
              child: Text('Cancel'),
              onPressed: () => Navigator.of(context).pop(),
            ),
            TextButton(
              child: Text('OK'),
              onPressed: () => Navigator.of(context).pop(),
            ),
          ],
        );
      });
}
```

```
```

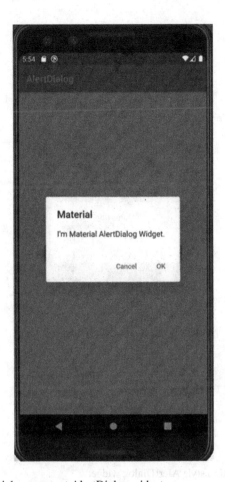

FIGURE 6.13   Material component AlertDialog widget

Figure 6.14 shows the Cupertino style AlertDialog widget.

SOURCE CODE ONLINE

The source code for this example (Flutter Widgets: AlertDialog) is available at GitHub.

CONCLUSION

In this chapter, you learned about some more Flutter widgets. We covered widgets to render images, toggle buttons, and text. You also learned about generating widgets asynchronously as the data becomes available, using FutureBuilder and StreamBuilder asynchronous widgets. Finally, we touched on Material and Cupertino style alert dialogs.

FIGURE 6.14   Cupertino style AlertDialog widget

REFERENCES

Flutter Dev. (2020, 11 18). *AlertDialog class.* Retrieved from Flutter Dev: https://api.flutter.
    dev/flutter/material/AlertDialog-class.html
Flutter Dev. (2020, 11 18). *Asynchronous programming: streams.* Retrieved from Dart Dev:
    https://dart.dev/tutorials/language/streams
Flutter Dev. (2020, 11 18). *ElevatedButton class.* Retrieved from Flutter Dev: https://api.
    flutter.dev/flutter/material/ElevatedButton-class.html
Flutter Dev. (2020, 11 18). *Future<T> class.* Retrieved from Flutter Dev: https://api.flutter.
    dev/flutter/dart-async/Future-class.html
Flutter Dev. (2020, 11 18). *FutureBuilder<T> class.* Retrieved from Flutter Dev: https://api.
    flutter.dev/flutter/widgets/FutureBuilder-class.html
Flutter Dev. (2020, 11 18). *Image class.* Retrieved from Flutter Dev: https://api.flutter.dev/
    flutter/widgets/Image-class.html
Flutter Dev. (2020, 11 18). *Placeholder class.* Retrieved from Flutter Dev: https://api.flutter.
    dev/flutter/widgets/Placeholder-class.html
Flutter Dev. (2020, 11 18). *Stream<T> class.* Retrieved from Dart Dev: https://api.dart.dev/
    stable/2.10.3/dart-async/Stream-class.html

Flutter Dev. (2020, 11 18). *StreamBuilder class*. Retrieved from Flutter Dev: https://api.flutter. dev/flutter/widgets/StreamBuilder-class.html

Flutter Dev. (2020, 11 18). *TextEditingController class*. Retrieved from Flutter Dev: https:// api.flutter.dev/flutter/widgets/TextEditingController-class.html

Flutter Dev. (2020, 11 18). *TextField class*. Retrieved from Flutter Dev: https://api.flutter.dev/ flutter/material/TextField-class.html

Flutter Dev. (2020, 11 18). *ToggleButtons class*. Retrieved from Flutter Dev: https://api.flutter. dev/flutter/material/ToggleButtons-class.html

Tyagi, P. (2020, 11 18). *Flutter Widgets: AlertDialog*. Retrieved from Chapter06: Pragmatic Flutter GitHub Repo: https://github.com/ptyagicodecamp/pragmatic_flutter/blob/ master/lib/chapter06/widgets/alert_dialog.dart

Tyagi, P. (2020, 11 18). *Flutter Widgets: FutureBuilder Async Widget*. Retrieved from Chapter06: Pragmatic Flutter GitHub Repo: https://github.com/ptyagicodecamp/ pragmatic_flutter/blob/master/lib/chapter06/widgets/futurebuilder_widget.dart

Tyagi, P. (2020, 11 18). *Flutter Widgets: Image*. Retrieved from Chapter06: Pragmatic Flutter GitHub Repo: https://github.com/ptyagicodecamp/pragmatic_flutter/blob/master/lib/ chapter06/widgets/image_widget.dart

Tyagi, P. (2020, 11 18). *Flutter Widgets: Placeholder*. Retrieved from Chapter06: Pragmatic Flutter GitHub Repo: https://github.com/ptyagicodecamp/pragmatic_flutter/blob/ master/lib/chapter06/widgets/placeholder_widget.dart

Tyagi, P. (2020, 11 18). *Flutter Widgets: StreamBuilder Async Widget*. Retrieved from Chapter06: Pragmatic Flutter GitHub Repo: https://github.com/ptyagicodecamp/ pragmatic_flutter/blob/master/lib/chapter06/widgets/streambuilder_widget.dart

Tyagi, P. (2020, 11 18). *Flutter Widgets: TextField*. Retrieved from Chapter06: Pragmatic Flutter GitHub Repo: https://github.com/ptyagicodecamp/pragmatic_flutter/blob/ master/lib/chapter06/widgets/textfield.dart

Tyagi, P. (2020, 11 18). *Flutter Widgets: ToggleButtons*. Retrieved from Chapter06: Pragmatic Flutter GitHub Repo: https://github.com/ptyagicodecamp/pragmatic_flutter/blob/ master/lib/chapter06/widgets/togglebuttons_widget.dart

Tyagi, P. (2021). Chapter 05: Flutter App Structure. In P. Tyagi, *Pragmatic Flutter: Building Cross-Platform Mobile Apps for Android, iOS, Web & Desktop*. CRC Press.

7 Building Layouts

In this chapter, you will learn to build layouts for a Material Flutter application. A Material app follows the material design guidelines (Material Components widgets). Flutter also supports building layouts for non-materials applications. You'll learn building layouts in Flutter (Layouts in Flutter) using layout Widgets (Layout widgets).

We will begin with revisiting the *HelloBooksApp* app from previous several chapters. This will help to understand the process of laying out widgets in a Flutter application.

REVISITING *HelloBooksApp* LAYOUT

Let's revisit the *HelloBooksApp* from the previous chapter (Chapter 05: Flutter App Structure) to analyze its four-part process of building its layout.

1. Choosing a layout widget
2. Creating a visible widget to display text
3. Adding visible widget to layout widget
4. Adding a layout widget to the app

Refer to Figure 7.1 to revisit the layout structure of *HelloBooksApp*.

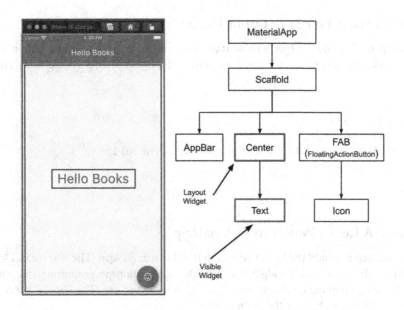

FIGURE 7.1 Layout of HelloBooksApp

The *HelloBooksApp* displays a greeting text in the middle of the screen. The smiley floating action button provides the option to switch the greeting to a different language on every click.

LAYOUT WIDGET

In the *HelloBooksApp*, the greeting text is displayed in the middle of the screen. We need a layout widget (Layout Widgets) to hold the visible widget displaying the text. The Center (Center class) layout widget centers its child horizontally and vertically inside it.

```
` ` `
Center (
)
` ` `
```

VISIBLE WIDGET

The Text widget displays the greeting text.

```
` ` `
Text (
    'Hello Books',
    style: Theme.of(context).textTheme.headline4,
),
` ` `
```

ADDING VISIBLE WIDGET TO LAYOUT WIDGET

The layout Center widget is more like a container. We need to add a visible widget as its descendant. At this point, we have got a layout with a visible text string.

```
` ` `
Center (
  child: Text (
    'Hello Books',
    style: Theme.of(context).textTheme.headline4,
  ),
)
` ` `
```

ADDING A LAYOUT WIDGET TO MaterialApp

The last step is to add the layout widget to the MaterialApp. The MaterialApp provides the Scaffold widget that provides application programming interfaces (APIs) to add app bars, bottom sheets, navigation drawer, etc. The `body` property adds the layout widget for the main screen.

```
```
MaterialApp(
  home: Scaffold(
    body: Center(
      child: Text(
        'Hello Books',
        style: Theme.of(context).textTheme.headline4,
      ),
    ),
  ),
);
```
```

In this section, you learned how to add layout widgets in a material Flutter app using the Center widget. In the rest of the chapter, you'll explore other layout widgets which are helpful in arranging widgets as per the app's design needs.

SOURCE CODE ONLINE

The source code for this example (Revisiting HelloBooksApp source code) is available at GitHub.

TYPES OF LAYOUT WIDGETS

There are two types of layout widgets:

1. Single child: There can be only one child added for these layouts. Few examples are: Center (Center class), Container (Container class), Padding (Padding class), ConstrainedBox (ConstrainedBox class), Expanded (Expanded class), Flexible (Flexible class), FittedBox (FittedBox class), FractionallySizedBox (FractionallySizedBox class), IntrinsicHeight (IntrinsicHeight class), IntrinsicWidth (IntrinsicWidth class), LimitedBox (LimitedBox class), OverflowBox (OverflowBox class), SizedOverflowBox (SizedOverflowBox class), SizedBox (SizedBox class), and Transform (Transform class).
2. Multi-child: Multiple children can be added to this type of layout widgets. Few examples are: Column (Column class), Row (Row class), Flex (Flex class), ListView (ListView class), GridView (GridView class), Stack (Stack class), IndexedStack (IndexedStack class), Flow (Flow class), LayoutBuilder (LayoutBuilder class), ListBody (ListBody class), Table (Table class), Wrap (Wrap class).

In the upcoming sections, we will be learning about some of the frequently used common layout widgets.

Container WIDGET

The Container (Container class) widget is a simple and versatile layout widget. It can hold only one child as its descendant. It has versatile properties that can be leveraged to style its child widget to change the background color or shape and size. By default, it aligns itself to the top-left corner of the screen.

Let's add a Text widget with 'Hello Container' text as its child. The TextStyle (Flutter Team, 2020) is applied to make the text's font '30'.

```
Container(
 child: Text(
   "Hello Container",
   style: TextStyle(fontSize: 30),
 ),
),
```

The 'Hello Container' will render like, as shown in Figure 7.2.

FIGURE 7.2   Container layout – default

You wouldn't notice any difference right after adding a widget as a child unless you use parameters/properties like `color`, `padding`, etc. Let's check out some of the `Container` widget's properties next.

Color PROPERTY

The `Container` layout widget's `color` property (color property) is used to highlight the background of the widget, as shown in Figure 7.3.

```
Container(
 color: Colors.red,
 child: Text(
   "Hello Container",
   style: TextStyle(fontSize: 30),
 ),
),
```

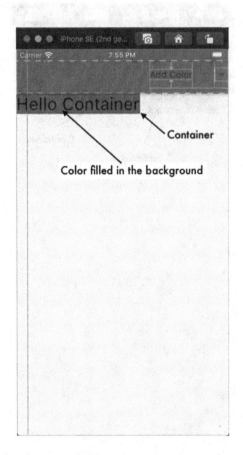

FIGURE 7.3   Container layout – color property

Padding PROPERTY

The padding property (padding property) adds empty space between the child widget and the Container (Container class) widget's boundary, as shown in Figure 7.4. The EdgeInsetsGeometry (EdgeInsetsGeometry class) is used to provide the padding of '16' points from all four sides.

```
~ ~ ~
Container(
 padding: const EdgeInsets.all(16.0),
 child: Text(
   "Hello Container",
   style: TextStyle(fontSize: 30),
 ),
),
~ ~ ~
```

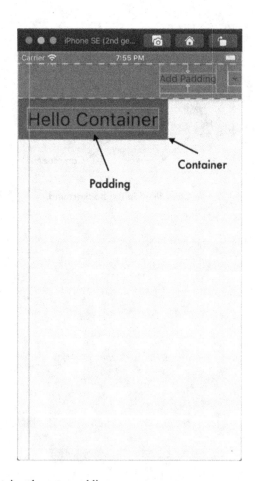

FIGURE 7.4 Container layout – padding property

Margin PROPERTY

The margin property (margin property) is used to add space surrounding the
Container widget, as shown in Figure 7.5.

```
```
Container(
 margin: const EdgeInsets.all(20.0),
 child: Text(
   "Hello Container",
   style: TextStyle(fontSize: 30),
 ),
),
```
```

Alignment PROPERTY

The `alignment` property (alignment property) is used to align the child within the
Container widget. The `Alignment.center` takes the parent widget's width

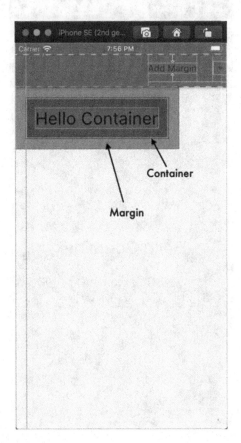

FIGURE 7.5 Container layout – margin property

and height, which can be constrained using the width and height property of the `Container` widget or using the `BoxConstraints` (BoxConstraints class). The widget expands to fill the parent's size.

```
Container(
 alignment: Alignment.center,
 child: Text(
   "Hello Container",
   style: TextStyle(fontSize: 30),
 ),
),
```

Refer to Figure 7.6 to observe alignment property of Container layout.

FIGURE 7.6 Container layout – alignment property

Constraints PROPERTY

The `constraints` property applies the size constraints to the Container widget. For example, BoxConstraints.tightFor(width:x, height:y) creates a box for the given width 'x', and/or height 'y'.

The Container widget looks like as shown in Figure 7.7 when width and height are constrained to be '100'. You can notice the text cut-off because the size of the box is not enough to display the full text.

```
```
Container(
 constraints: BoxConstraints.tightFor(width: 100.0, height:
100.0),
  child: Text(
    "Hello Container",
    style: TextStyle(fontSize: 30),
  ),
),
```
```

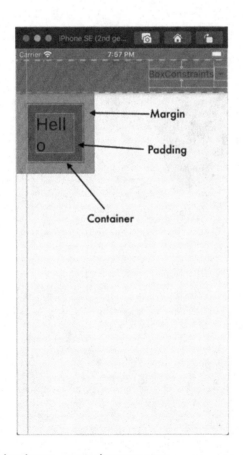

FIGURE 7.7   Container layout – constraints property

Transform PROPERTY

The `transform` property (transform property) is used to transform the child before adding to the layout widget `Container`. The value `Matrix4.rotationZ(0.3)` rotates the Container widget clockwise by the given amount.

```
```
Container(
 transform: Matrix4.rotationZ(0.3)
 child: Text(
   "Hello Container",
   style: TextStyle(fontSize: 30),
 ),
),
```
```

Refer to Figure 7.8 to observe transform property of Container layout.

FIGURE 7.8   Container layout – transform property

Decoration PROPERTY

The `decoration` property is used to add shape to the Container widget. It uses BoxDecoration to provide details. The BoxDecoration (BoxDecoration class) tells the widget how the box around the Container widget will be painted. In simple words, this property lets Container create a border around it or drop a shadow. Let's create a solid border around the Container of specified width and color. Please note that the Container widget's `color` property cannot be used along with `decoration`. The Container widget's `color` and `decoration` property can't be used together.

```
```
Container(
  decoration: BoxDecoration(
    border: Border.all(
      color: Colors.amber,
      width: 5.0,
      style: BorderStyle.solid,
    ),
  ),
  child: Text(
    "Hello Container",
    style: TextStyle(fontSize: 30),
  ),
),
```
```

Refer to Figure 7.9 to observe decoration property in Container layout.

SOURCE CODE ONLINE

The source code for this example (Building Layouts: Container Widget) is available online at GitHub.

Padding WIDGET

The Padding (Padding class) widget insets its child as per the given padding. It creates empty space around the child. It takes care of resizing any constraints passed to the child to be able to provide the given empty space around it.

A Padding widget is created as below. It uses the `padding` property to assign the amount of space for the inset. The EdgeInsets (EdgeInsets class) class specifies the offset from all four edges. In the example below, an offset of 8 dip (Density-independent Pixels) is provided from the top, bottom, left, and right sides. The keyword `const` is required when you are sure that the provided padding won't change. In this example, we'll modify padding to show a few examples of different values for padding property.

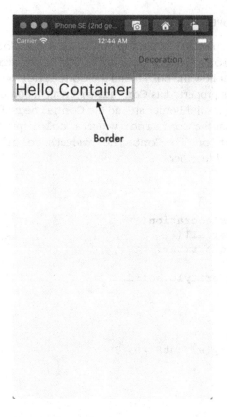

FIGURE 7.9   Container layout – decoration property

```
double padding = 8.0;
Padding(
 padding: EdgeInsets.all(padding),
 child: Text(
   "Hello Padding",
   style: TextStyle(fontSize: 30),
 ),
)
```

Figure 7.10 shows the Padding widget- a single child layout.

The difference between `padding` and `margin` is that Padding creates the empty space around the child widget of the layout widget like Container. The `margin` property creates the space around the layout widget itself.

SOURCE CODE ONLINE

The source code for this example (Building Layouts: Padding Widget) is available online at GitHub.

Padding defaulted to 8.0 pixels from all sides.

Padding increased to 104.0 pixels from all sides.

Padding decreased to 88.0 pixels from all sides.

FIGURE 7.10   Padding single-child layout widget

`ConstrainedBox` WIDGET

Sometimes you may want to render a widget of a given size. The `ConstrainedBox` (ConstrainedBox class) is a layout widget that puts additional constraints on its child. Let's check out three types of `BoxContraints` applied to the `ConstrainedBox` widget. It specifies a maximum and minimum width and height its child is allowed to expand. `BoxConstraints.expand()` fills the parent.

MINIMUM WIDTH & HEIGHT

This constraint imposes minimum width and height on the child. In this example, a `ConstrainedBox` widget is added in the center of the body of the app. A `Container` widget is added as the child displaying a message in the `Text` widget.

```
```
...
body: Center(
  child: ConstrainedBox(
    constraints: BoxConstraints(
      minWidth: 100,
      minHeight: 100,
    ),
    child: Container(
      color: Colors.grey,
      child: Text(message),
```
```

```
    ),
  ),
),
...
```

BoxConstraints.expand()

The `BoxConstraints.expand()` let its child expand to the given width and height.

```
...
body: Center(
  child: ConstrainedBox(
    constraints: BoxConstraints.expand(
      width: 200,
      height: 200,
    ),
    child: Container(
      color: Colors.grey,
      child: Text(message),
    ),
  ),
),
...
```

BoxConstraints.loose()

The `BoxConstraints.loose()` constrains its child to the given size. It can't go beyond the provided size.

```
...
body: Center(
  child: ConstrainedBox(
    constraints: BoxConstraints.loose(
      Size(100, 200),
    ),
    child: Container(
      color: Colors.grey,
      child: Text(message),
    ),
  ),
),
...
```

All three constraints are shown side by side in Figure 7.11.

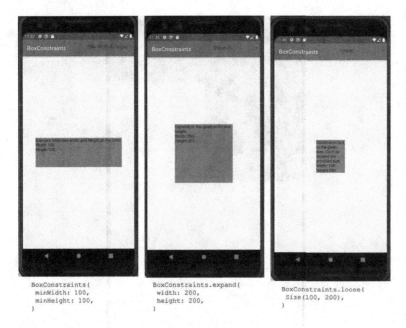

BoxConstraints(
    minWidth: 100,
    minHeight: 100,
)

BoxConstraints.expand(
    width: 200,
    height: 200,
)

BoxConstraints.loose(
    Size(100, 200),
)

FIGURE 7.11   ConstrainedBox Widget – constraints properties

SOURCE CODE ONLINE

The source code for this example (Building Layouts: ConstrainedBox Widget) is available online at GitHub.

SizedBox WIDGET

The SizedBox (SizedBox class) widget is a Single-child layout widget. It's a box widget of a specific size and can add one another widget as its child. It's useful when you know the size of the widget. The `width` property is used to set the width of the box, and the `height` property is used to set the box's height.

The code snippet below creates a `SizedBox` of height and width of 200 device-independent pixels (devicePixelRatio property).

```
SizedBox(
  height: 200,
  width: 200,
  child: Container(
    color: Colors.deepPurpleAccent,
  ),
)
```

Figure 7.12 displays SizedBox widget.

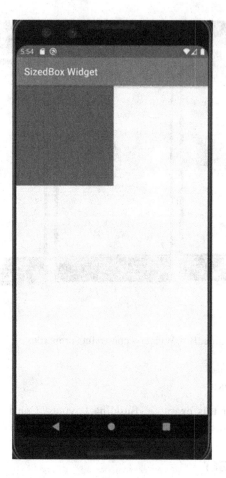

FIGURE 7.12   SizedBox with height and width as 200 dips

There's a convenience constructor, `SizedBox.expand` (SizedBox.expand constructor), that can also be used to create a box that takes the width and height of its parent.

```
```
SizedBox.expand(
    child: Container(
      color: Colors.deepPurpleAccent,
    ),
)
```
```

Figure 7.13 demonstrates usage of SizedBox.expand constructor to create SizedBox widget.

The same results can be attained by assigning the `SizedBox` widget's `width` and `height` properties to `double.infinity`.

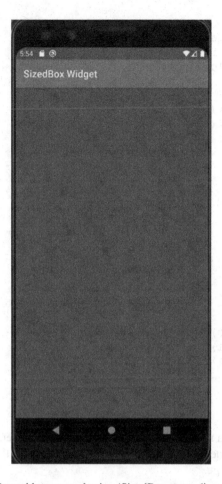

FIGURE 7.13   SizedBox widget created using 'SizedBox.expand' constructor

It is common to use a `SizedBox` without a child to add the space between widgets when building interfaces.

SOURCE CODE ONLINE

The source code for this example (Building Layouts: SizedBox Widget) is available online at GitHub.

Row WIDGET

The Row (Row class) widget is used to arrange its children in a horizontal fashion. Let's add three Container widgets as children. The `childWidget(int index)` method returns a Container widget of width and height device-independent pixels or dips (devicePixelRatio property). The container has a Text widget as its child, which displays the number passed to the method as a parameter.

```
` ` `
Widget childWidget(int index) {
  return Container(
    color: getColor(index),
    width: 100,
    height: 100,
    child: Center(
      child: Text(
        "$index",
        style: TextStyle(fontSize: 40),
      ),
    ),
  );
}
` ` `
```

Now that we have got a child, let's add this three times in the Row widget as below:

```
` ` `
Row(
  children: [
    childWidget(0),
    childWidget(1),
    childWidget(2),
  ],
),
` ` `
```

Figure 7.14 displays a multi-child layout built with Row widget.

Let's try to add one more widget, `childWidget(3)` to Row widget's children and observe the change.

```
` ` `
Row(
  children: [
    childWidget(0),
    childWidget(1),
    childWidget(2),
    childWidget(3),
  ],
),
` ` `
```

You'll notice that there's not enough space for all four widgets to fit horizontally. The last child renders with yellow and black lines. You will see those lines whenever a widget overflows the available space to render itself. You'll learn about creating adaptable layouts in the chapter on building responsive layouts (Chapter 08: Responsive Interfaces).

Figure 7.15 shows the overflowing widgets in a multi-child layout.

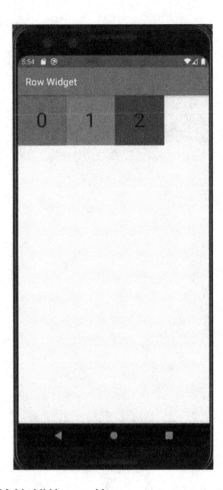

FIGURE 7.14   Row: Multi-child layout widget

Source Code Online

The source code for this example (Building Layouts: Row Widget) is available online at GitHub.

IntrinsicHeight WIDGET

The `IntrinsicHeight` (IntrinsicHeight class) widget is a Single-child layout widget. `IntrinsicHeight` widget helps to set the height of its child widget when there's unlimited height available to it. This class is expensive. The cheap way of limiting the widget size is to use the `SizedBox` (SizedBox class) layout widget. This widget is used when children of a `Row` (Row class) widget are expected to expand to the height of the tallest child.

Let's understand this with the help of an example. First, create children of varying sizes with the help of the following `childWidget(int index)` method:

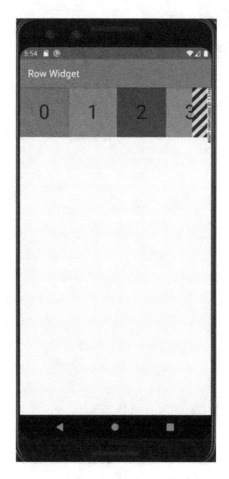

FIGURE 7.15   Row: overflowing child

```
Widget childWidget(int index) {
 return Container(
   color: getColor(index),
   width: 100 + index * 20.toDouble(),
   height: 100 + index * 30.toDouble(),
   child: Center(
     child: Text(
       "$index",
       style: TextStyle(fontSize: 40),
     ),
   ),
 );
}
```

Next, add three children created by the `childWidget()` method to the Row widget as below:

```
Row (
 children: [
   childWidget(0),
   childWidget(1),
   childWidget(2),
 ],
),
```

The code above will render the three Container widgets of varying sizes in a horizontal array, as shown in Figure 7.16.

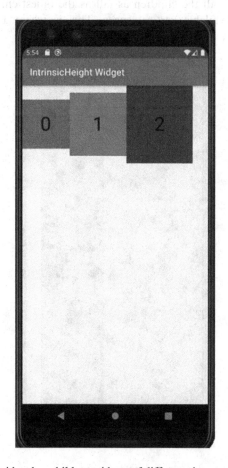

FIGURE 7.16   Row widget has children widgets of different sizes

The goal is to stretch all the children to the same height. So, let's set the Row widget's `crossAxisAlignment` property to `CrossAxisAlignment.stretch` to make all children equally tall.

```
```

Row(
  crossAxisAlignment: CrossAxisAlignment.stretch,
  children: [
    childWidget(0),
    childWidget(1),
    childWidget(2),
  ],
),
```
```

However, the problem is that they'll take up all the available space vertically and make it look like, as shown in Figure 7.17.

We want to make all the children as tall as the tallest child widget while not taking up all the available vertical space. This is where IntrinsicHeight

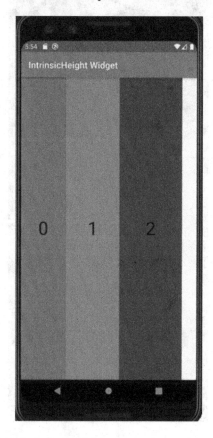

FIGURE 7.17   Row widget's 'crossAxisAlignment' property is set to 'CrossAxisAlignment. stretch'

(IntrinsicHeight class) widget comes to play. All you need to do is to wrap the Row widget inside `IntrinsicHeight,` as shown in the code snippet below:

```
```
IntrinsicHeight(
 child: Row(
   crossAxisAlignment: CrossAxisAlignment.stretch,
   children: [
     childWidget(0),
     childWidget(1),
     childWidget(2),
   ],
 ),
),
```
```

The `IntrinsicHeight` widget expands all of the Row widget's children to the same height as the tallest child widget as shown in Figure 7.18.

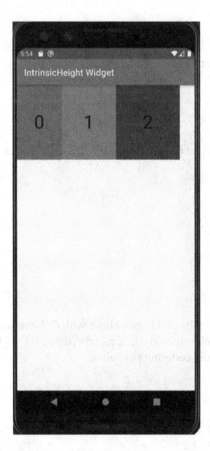

FIGURE 7.18 'IntrinsicHeight' widget expands all of the Row widget's children to the same height as the tallest child widget

The source code for this example (Building Layouts: IntrinsicHeight Widget) is available online at GitHub.

Column WIDGET

The Column (Column class) widget is used to arrange its children in a vertical manner. Let's add three Container widgets as children similar to the Row widget. The `childWidget(int index)` method returns a Container widget of width and height as 200 dips. The container has a Text widget as its child, which displays the number passed to the method as a parameter.

```
Container(
  color: getColor(index),
  width: 200,
  height: 200,
  child: Center(
    child: Text(
      "$index",
      style: TextStyle(fontSize: 40),
    ),
  ),
)
```

Let's add this child widget three times in the Column widget as below:

```
Column(
  children: [
    childWidget(0),
    childWidget(1),
    childWidget(2),
  ],
),
```

Figure 7.19 displays a multi-child layout built with Column widget.

Let's try to add one more widget, `childWidget(3)` to the Column widget's children, as shown in the code snippet below:

```
Column(
  children: [
    childWidget(0),
    childWidget(1),
    childWidget(2),
```

```
  childWidget(3),
 ],
),
```

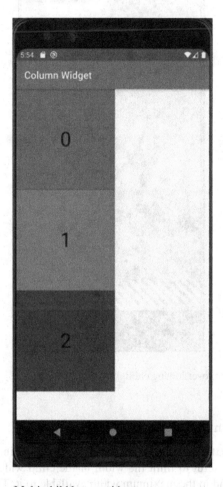

FIGURE 7.19   Column: Multi-child layout widget

You'll notice the same yellow and black overflow lines that you observed earlier in the Row widget. This is because there's not enough space for all four widgets to fit vertically as shown in Figure 7.20.

SOURCE CODE ONLINE

The source code for this example (Building Layouts: Column Widget) is available online at GitHub.

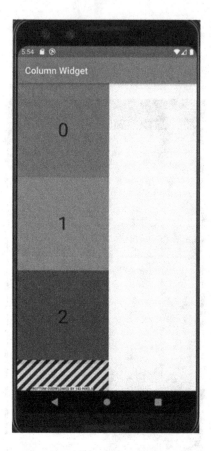

FIGURE 7.20   Column: overflowing child

IntrinsicWidth WIDGET

The IntrinsicWidth (IntrinsicWidth class) widget is a Single-child layout widget. This widget is useful to limit the width of the child widget to a given width; otherwise, it'll expand to the maximum width available to it. This widget is usually used when each child of a Column widget is expected to have the same width. All children expand to the width of the widest child widget of the Column widget.

Let's use the asymmetric Container widgets generated by the `childWidget(int index)` method used in the IntrinsicHeight section. Add three of these child widgets to the Column widget as below:

```
```

```
Column(
 children: [
    childWidget(0),
    childWidget(1),
    childWidget(2),
```

```
],
),
~ ~ ~
```

The three children with different sizes will be shown in Column widgets, as shown in Figure 7.21.

We can set the `crossAxisAlignment` property to `CrossAxisAlignment.stretch` to make all Container children of the same widths.

```
~ ~ ~
Column(
  crossAxisAlignment: CrossAxisAlignment.stretch,
  children: [
    childWidget(0),
    childWidget(1),
    childWidget(2),
  ],
),
~ ~ ~
```

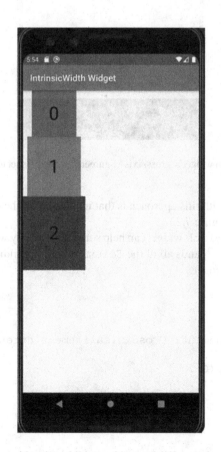

FIGURE 7.21   Column widget has children widgets of different sizes

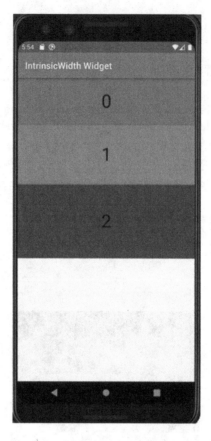

FIGURE 7.22   Column widget's 'crossAxisAlignment' property is set to 'CrossAxisAlignment. stretch'

Again, the problem with this approach is that it takes up all the cross-axis horizontal space, as shown in Figure 7.22.

The `IntrinsicWidth` widget can help solve this issue by wrapping the `Column` widget as its child. It expands all of the `Column` widget's children to the same width as the widest child widget.

```
IntrinsicWidth(
 child: Column(
   crossAxisAlignment: CrossAxisAlignment.stretch,
   children: [
     childWidget(0),
     childWidget(1),
     childWidget(2),
   ],
 ),
),
```

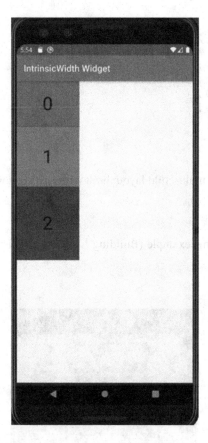

FIGURE 7.23   'IntrinsicWidth' widget expands all of the column widget's children to the same width as the widest child widget

Figure 7.23 demonstrates IntrinsicWidth widget.

SOURCE CODE ONLINE

The source code for this example (Building Layouts: IntrinsicWidth Widget) is available online at GitHub.

ListView WIDGET

The ListView (ListView class) widget is a Multi-child and scrolling widget. It makes its children scroll in the main axis while filling the space in the cross-axis. Let's add four children to the ListView widget, as shown in the code snippet below. We're using the same method `childWidget(int index)` to create the child widget(s) as we did for the Column widget in the previous section.

```
` ` `
ListView(
  children: [
    childWidget(0),
    childWidget(1),
    childWidget(2),
    childWidget(3),
  ],
),
` ` `
```

Figure 7.24 displays a multi-child layout built with ListView widget.

SOURCE CODE ONLINE

The source code for this example (Building Layouts: ListView Widget) is available online at GitHub.

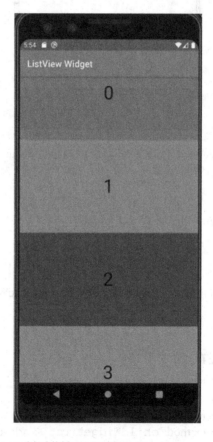

FIGURE 7.24   ListView multi-child layout widget

`GridView` WIDGET

The `GridView` (GridView class) widget is a Multi-child and scrolling widget like `ListView`. It arranges its children in a two-dimensional array. In this section, we will focus on creating a grid layout using `GridView.count` (GridView.count constructor) constructor. It creates a grid with a given number of tiles on the cross-axis. The direction a `GridView` scroll is the main-axis. The `crossAxisCount` property is used for the number of tiles arranged in the cross-axis. In the following code snippet, `crossAxisCount` is two, which means there are two tiles in the horizontal direction.

```
GridView.count(
 crossAxisCount: 2,
 children: [
   childWidget(0),
   childWidget(1),
   childWidget(2),
   childWidget(3),
 ],
),
```

Figure 7.25 displays a multi-child layout built with GridView widget.

SOURCE CODE ONLINE

The source code for this example (Building Layouts: GridView Widget) is available online at GitHub.

`Table` WIDGET

The `Table` (Table class) layout widget supports multiple children. This layout widget arranges its children in a tabular layout. This layout doesn't scroll and is used for a fixed number of widgets. The `Table` layout widget is useful to design interfaces that don't require any scrolling and to avoid multiple levels of nested `Row` and `Column` widgets. The `Table` widget can be wrapped inside a `SingleChildScrollView` (SingleChildScrollView class) to make it scrollable. The `SingleChildScrollView` widget is like a scrollable box, which makes its only child scrollable.

Let's add four children to the `Table` layout. Each child is a `Container` widget of different sizes and colors. The `childWidget()` method is used to create such children widgets.

```
Widget childWidget(int index) {
 return Container(
   color: getColor(index),
```

```
    width: 200 + index * 20.toDouble(),
    height: 200 + index * 30.toDouble(),
    child: Center(
      child: Text(
        "$index",
        style: TextStyle(fontSize: 40),
      ),
    ),
  );
}
```

The `Table` widget has multiple children of type `TableRow` widget, as shown in the code snippet below. The `border` property is used to border each cell. The `column-nWidths` property is used to assign the width for the given column. It's a mapping between the column number to `FractionColumnWidth` (FractionColumnWidth class) for the given column.

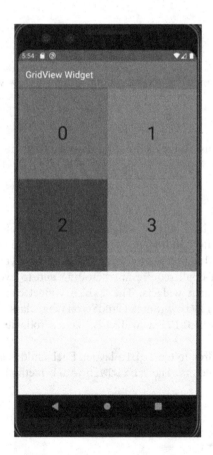

FIGURE 7.25   GridView multi-child layout widget

```
```
Table(
  border: TableBorder.all(width: 2.0),
  columnWidths: {
    0: FractionColumnWidth(.5),
    1: FractionColumnWidth(.5),
  },
  children: [
    TableRow(
      children: [
        childWidget(0),
        childWidget(1),
      ],
    ),
    TableRow(
      children: [
        childWidget(2),
        childWidget(3),
      ],
    ),
  ],
)
```
```

Figure 7.26 displays a multi-child layout built with Table widget.

SOURCE CODE ONLINE

The source code for this example (Building Layouts: Table Widget) is available online at GitHub.

Stack WIDGET

The Stack (Stack class) widget is a Multi-child layout widget since it can hold multiple children. It can stack one widget on top of another widget, just as its name says. This widget is useful when one widget is required to be overlapped by another. Let's take an example to understand the usage of Stack widgets. We will create three widgets of varying sizes using the `childWidget(int index)` method. This method takes an integer and creates and returns a Container widget. Its width and height are calculated using the `index` parameter. The function `getColor(int index)` returns different colors based on the `index` parameter.

```
```
Widget childWidget(int index) {
  return Container(
    color: getColor(index),
    width: 200 + index * 20.toDouble(),
    height: 200 + index * 30.toDouble(),
    child: Center(
```
```

```
        child: Text(
          "$index",
          style: TextStyle(fontSize: 40),
        ),
      ),
    );
}
```

Next, put three `childWidgets()` in the Stack widget as below:

```
Stack(
 children: [
   childWidget(2),
   childWidget(1),
   childWidget(0),
 ],
),
```

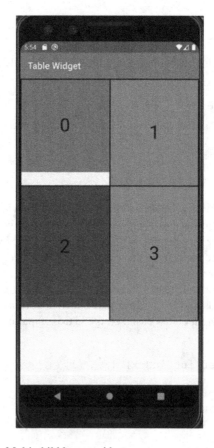

FIGURE 7.26   Table – Multi-child layout widget

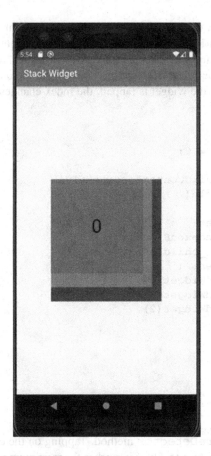

FIGURE 7.27   Stack – Multi-child layout widget

The `childWidget(2)` is the Container widget with the biggest width and height, and it's purple in color. This widget goes first. The next child widget, `childWidget(1)`, is green and slightly smaller than the purple widget. It is placed on top of the purple widget. The third widget, `childWidget(0)`, is red and smallest. It's placed on top of the stack. The stack looks like as shown in Figure 7.27.

SOURCE CODE ONLINE

The source code for this example (Building Layouts: Stack Widget) is available online at GitHub.

IndexedStack WIDGET

The IndexedStack (IndexedStack class) widget is a Multi-child layout widget as well. It's like the Stack widget but only shows one child at a time. It uses `index` property to switch from one child to another. Let's use the same example

we discussed above for the Stack widget. This time these three `childWidget()`
are wrapped inside IndexedStack instead of Stack widget and have an `index`
property to show the currently selected child. We will wrap IndexedStack inside
a GestureDetector (GestureDetector class) widget to add a tapping gesture on
the widget. Each time the widget is tapped, the index changes to the next child and
keeps going one by one.

```
int _childIndex = 0;
@override
Widget build(BuildContext context) {
  return Scaffold(
    ...,
    body: Center(
      child: IndexedStack(
        index: _childIndex,
        children: [
          childWidget(0),
          childWidget(1),
          childWidget(2),
        ],
      ),
    ),
  );
}
```

The `childWidget(int index)` method returns a Container widget
wrapped in a GestureDetector method. Tapping on the child widget increases
the `_childIndex` by one. Once it reaches the maximum possible index two, the
`_childIndex` resets itself to zero. The clamp (clamp method) method returns
the number between zero and two.

```
Widget childWidget(int index) {
  return GestureDetector(
    onTap: () {
      setState(() {
        index = index == 2? 0 : index + 1;
        _childIndex = index.clamp(0, 2);
      });
    },
    child: Container(
      color: getColor(index),
      width: 200 + index * 20.toDouble(),
      height: 200 + index * 30.toDouble(),
      child: Center(
        child: Text(
          "$index",
          style: TextStyle(fontSize: 40),
```

```
        ),
       ),
      ),
     );
   }
```

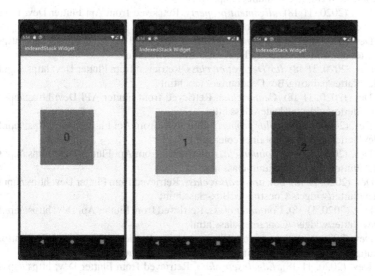

FIGURE 7.28   IndexedStack – Multi-child layout widget

The red widget is stacked at the top, followed by green and purple. Ordering of children doesn't matter in the case of the `IndexedStack` widget. The child widget is displayed based on the selected `index` property.

Figure 7.28 displays a multi-child layout built with IndexedStack widget.

Source Code Online

The source code for this example (Building Layouts: IndexedStack Widget) is available online at GitHub.

CONCLUSION

In this chapter, Flutter layout widgets were covered and discussed. We learned that there are two types of layout widgets: Single- and Multi-child layouts. The Single-child layout can have one child whereas Multi-child layout can have multiple children. We revisited the *HelloBooksApp* to understand its layout structure. We discussed Single-child layouts like `Container`, `Padding`, `ConstrainedBox`, `SizedBox`, `IntrinsicHeight`, and `IntrinsicWidth`. You also learned about the Multi-child layouts like `Row`, `Column`, `ListView`, `GridView`, `Table`, `Stack`, and `IndexedStack`.

REFERENCES

Android Developer. (2020, 12 20). *Density-independent Pixels.* Retrieved from developer. android.com: https://developer.android.com/guide/topics/resources/more-resources. html#Dimension

Dart Dev. (2020, 11 18). *clamp method.* Retrieved from Dart Dev: https://api.dart.dev/ stable/2.10.2/dart-core/num/clamp.html

Flutter Dev. (2020, 11 18). *alignment property.* Retrieved from Api Flutter Dev: https://api. flutter.dev/flutter/widgets/Container/alignment.html

Flutter Dev. (2020, 11 18). *BoxConstraints class.* Retrieved from Api Flutter Dev: https://api. flutter.dev/flutter/rendering/BoxConstraints-class.html

Flutter Dev. (2020, 11 18). *BoxDecoration class.* Retrieved from Flutter Dev: https://api.flutter. dev/flutter/painting/BoxDecoration-class.html

Flutter Dev. (2020, 11 18). *Center class.* Retrieved from Flutter API Dev: https://api.flutter. dev/flutter/widgets/Center-class.html

Flutter Dev. (2020, 11 18). *color property.* Retrieved from Api Flutter Dev: https://api.flutter. dev/flutter/widgets/Container/color.html

Flutter Dev. (2020, 11 18). *Column class.* Retrieved from Api Flutter Dev: https://api.flutter. dev/flutter/widgets/Column-class.html

Flutter Dev. (2020, 11 18). *ConstrainedBox class.* Retrieved from Flutter Dev: https://api.flutter. dev/flutter/widgets/ConstrainedBox-class.html

Flutter Dev. (2020, 11 18). *Container class.* Retrieved from Flutter Api Dev: https://api.flutter. dev/flutter/widgets/Container-class.html

Flutter Dev. (2020, 11 18). *devicePixelRatio property.* Retrieved from Flutter Dev: https://api. flutter.dev/flutter/dart-ui/Window/devicePixelRatio.html

Flutter Dev. (2020, 11 18). *EdgeInsets class.* Retrieved from Flutter Dev: https://api.flutter. dev/flutter/painting/EdgeInsets-class.html

Flutter Dev. (2020, 11 18). *EdgeInsetsGeometry class.* Retrieved from Api Flutter Dev: https://api.flutter.dev/flutter/painting/EdgeInsetsGeometry-class.html

Flutter Dev. (2020, 11 18). *Expanded class.* Retrieved from Flutter Dev: https://api.flutter.dev/ flutter/widgets/Expanded-class.html

Flutter Dev. (2020, 11 18). *FittedBox class.* Retrieved from Api Flutter Dev: https://api.flutter. dev/flutter/widgets/FittedBox-class.html

Flutter Dev. (2020, 11 18). *Flex class.* Retrieved from Api Flutter Dev: https://api.flutter.dev/ flutter/widgets/Flex-class.html

Flutter Dev. (2020, 11 18). *Flexible class.* Retrieved from Api Flutter Dev: https://api.flutter. dev/flutter/widgets/Flexible-class.html

Flutter Dev. (2020, 11 18). *Flow class.* Retrieved from Flutter Dev: https://api.flutter.dev/flutter/ widgets/Flow-class.html

Flutter Dev. (2020, 11 18). *FractionallySizedBox class.* Retrieved from Api Flutter Dev: https://api.flutter.dev/flutter/widgets/FractionallySizedBox-class.html

Flutter Dev. (2020, 11 18). *FractionColumnWidth class.* Retrieved from Flutter Dev: https:// api.flutter.dev/flutter/rendering/FractionColumnWidth-class.html

Flutter Dev. (2020, 11 18). *GestureDetector class.* Retrieved from Flutter Dev: https://api. flutter.dev/flutter/widgets/GestureDetector-class.html

Flutter Dev. (2020, 11 18). *GridView class.* Retrieved from Api Flutter Dev: https://api.flutter. dev/flutter/widgets/GridView-class.html

Flutter Dev. (2020, 11 18). *GridView.count constructor.* Retrieved from Flutter Dev: https:// api.flutter.dev/flutter/widgets/GridView/GridView.count.html

Flutter Dev. (2020, 11 18). *IndexedStack class.* Retrieved from Flutter Dev: https://api.flutter. dev/flutter/widgets/IndexedStack-class.html

Flutter Dev. (2020, 11 18). *IntrinsicHeight class*. Retrieved from Api Flutter Dev: https://api.
flutter.dev/flutter/widgets/IntrinsicHeight-class.html
Flutter Dev. (2020, 11 18). *IntrinsicWidth class*. Retrieved from Api Flutter Dev: https://api.
flutter.dev/flutter/widgets/IntrinsicWidth-class.html
Flutter Dev. (2020, 11 18). *Layout Widgets*. Retrieved from Flutter Dev: https://flutter.dev/
docs/development/ui/widgets/layout
Flutter Dev. (2020, 11 18). *LayoutBuilder class*. Retrieved from Flutter Team: https://api.
flutter.dev/flutter/widgets/LayoutBuilder-class.html
Flutter Dev. (2020, 11 18). *LimitedBox class*. Retrieved from Api Flutter Dev: https://api.
flutter.dev/flutter/widgets/LimitedBox-class.html
Flutter Dev. (2020, 11 18). *ListBody class*. Retrieved from Flutter Dev: https://api.flutter.dev/
flutter/widgets/ListBody-class.html
Flutter Dev. (2020, 11 18). *ListView class*. Retrieved from Api Flutter Dev: https://api.flutter.
dev/flutter/widgets/ListView-class.html
Flutter Dev. (2020, 11 18). *margin property*. Retrieved from Api Flutter Dev: https://api.
flutter.dev/flutter/widgets/Container/margin.html
Flutter Dev. (2020, 12 18). *Material Components widgets*. Retrieved from Flutter Dev: https://
flutter.dev/docs/development/ui/widgets/material
Flutter Dev. (2020, 11 18). *OverflowBox class*. Retrieved from Api Flutter Dev: https://api.
flutter.dev/flutter/widgets/OverflowBox-class.html
Flutter Dev. (2020, 11 18). *Padding class*. Retrieved from Flutter Api Dev: https://api.flutter.
dev/flutter/widgets/Padding-class.html
Flutter Dev. (2020, 11 18). *padding property*. Retrieved from Flutter Dev: https://api.flutter.
dev/flutter/widgets/Container/padding.html
Flutter Dev. (2020, 11 18). *Row class*. Retrieved from Api Flutter Dev: https://api.flutter.dev/
flutter/widgets/Row-class.html
Flutter Dev. (2020, 11 18). *SizedBox class*. Retrieved from Api Flutter Dev: https://api.flutter.
dev/flutter/widgets/SizedBox-class.html
Flutter Dev. (2020, 11 18). *SizedBox.expand constructor*. Retrieved from Flutter Dev: https://
api.flutter.dev/flutter/widgets/SizedBox/SizedBox.expand.html
Flutter Dev. (2020, 11 18). *SizedOverflowBox class*. Retrieved from Api Flutter Dev: https://
api.flutter.dev/flutter/widgets/SizedOverflowBox-class.html
Flutter Dev. (2020, 11 18). *Stack class*. Retrieved from Api Flutter Dev: https://api.flutter.dev/
flutter/widgets/Stack-class.html
Flutter Dev. (2020, 11 18). *Table class*. Retrieved from Flutter Dev: https://api.flutter.dev/
flutter/widgets/Table-class.html
Flutter Dev. (2020, 11 18). *Transform class*. Retrieved from Api Flutter Dev: https://api.flutter.
dev/flutter/widgets/Transform-class.html
Flutter Dev. (2020, 11 18). *transform property*. Retrieved from Api Flutter Dev: https://api.
flutter.dev/flutter/widgets/Container/transform.html
Flutter Dev. (2020, 11 18). *Wrap class*. Retrieved from Api Flutter Dev: https://api.flutter.dev/
flutter/widgets/Wrap-class.html
Flutter Team. (2020, 11 18). *SingleChildScrollView class*. Retrieved from Flutter Dev: https://
api.flutter.dev/flutter/widgets/SingleChildScrollView-class.html
Flutter Team. (2020, 11 18). *TextStyle class*. Retrieved from Flutter Dev: https://api.flutter.
dev/flutter/painting/TextStyle-class.html
Google. (2020, 11 18). *Layout widgets*. Retrieved from Flutter Dev: https://flutter.dev/docs/
development/ui/widgets/layout
Google. (2020, 11 18). *Layouts in Flutter*. Retrieved from Flutter Dev: https://flutter.dev/docs/
development/ui/layout

Tyagi, P. (2020, 11 18). *Building Layouts: Column Widget*. Retrieved from Chapter07: Pragmatic Flutter GitHub Repo: https://github.com/ptyagicodecamp/pragmatic_flutter/ blob/master/lib/chapter07/layouts/multi/column.dart

Tyagi, P. (2020, 11 18). *Building Layouts: ConstrainedBox Widget*. Retrieved from Chapter07: Pragmatic Flutter GitHub Repo: https://github.com/ptyagicodecamp/pragmatic_flutter/ blob/master/lib/chapter07/layouts/single/constrained_box.dart#L109:L113

Tyagi, P. (2020, 11 18). *Building Layouts: Container Widget*. Retrieved from Chapter07: Pragmatic Flutter GitHub Repo: https://github.com/ptyagicodecamp/pragmatic_flutter/ blob/master/lib/chapter07/layouts/single/container.dart

Tyagi, P. (2020, 11 18). *Building Layouts: GridView Widget*. Retrieved from Chapter07: Pragmatic Flutter GitHub Repo: https://github.com/ptyagicodecamp/pragmatic_flutter/ blob/master/lib/chapter07/layouts/multi/grid_view.dart

Tyagi, P. (2020, 11 18). *Building Layouts: IndexedStack Widget*. Retrieved from Chapter07: Pragmatic Flutter GitHub Repo: https://github.com/ptyagicodecamp/pragmatic_flutter/ blob/master/lib/chapter07/layouts/multi/indexed_stack.dart

Tyagi, P. (2020, 11 18). *Building Layouts: IntrinsicHeight Widget*. Retrieved from Chapter07: Pragmatic Flutter GitHub Repo: https://github.com/ptyagicodecamp/pragmatic_flutter/ blob/master/lib/chapter07/layouts/single/intrinsic_height.dart

Tyagi, P. (2020, 11 18). *Building Layouts: IntrinsicWidth Widget*. Retrieved from Chapter07: Pragmatic Flutter GitHub Repo: https://github.com/ptyagicodecamp/pragmatic_flutter/ blob/master/lib/chapter07/layouts/single/intrinsic_width.dart

Tyagi, P. (2020, 11 18). *Building Layouts: ListView Widget*. Retrieved from Chapter07: Pragmatic Flutter GitHub Repo: https://github.com/ptyagicodecamp/pragmatic_flutter/ blob/master/lib/chapter07/layouts/multi/listview.dart

Tyagi, P. (2020, 11 18). *Building Layouts: Padding Widget*. Retrieved from Chapter07: Pragmatic Flutter GitHub Repo: https://github.com/ptyagicodecamp/pragmatic_flutter/ blob/master/lib/chapter07/layouts/single/padding.dart#L92:L98

Tyagi, P. (2020, 11 18). *Building Layouts: Row Widget*. Retrieved from Chapter07: Pragmatic Flutter GitHub Repo: https://github.com/ptyagicodecamp/pragmatic_flutter/blob/master/lib/chapter07/layouts/multi/row.dart

Tyagi, P. (2020, 11 18). *Building Layouts: SizedBox Widget*. Retrieved from Chapter07: Pragmatic Flutter GitHub Repo: https://github.com/ptyagicodecamp/pragmatic_flutter/ blob/master/lib/chapter07/layouts/single/sized_box.dart

Tyagi, P. (2020, 11 18). *Building Layouts: Stack Widget*. Retrieved from Chapter07: Pragmatic Flutter GitHub Repo: https://github.com/ptyagicodecamp/pragmatic_flutter/blob/master/lib/chapter07/layouts/multi/stack.dart

Tyagi, P. (2020, 11 18). *Building Layouts: Table Widget*. Retrieved from Chapter07: Pragmatic Flutter GitHub Repo: https://github.com/ptyagicodecamp/pragmatic_flutter/blob/master/lib/chapter07/layouts/multi/table.dart

Tyagi, P. (2020, 11 18). *Revisiting HelloBooksApp source code*. Retrieved from Chapter05: Pragmatic Flutter GitHub Repo: https://github.com/ptyagicodecamp/pragmatic_flutter/ blob/master/lib/chapter05/main_05_3.dart

Tyagi, P. (2021). Chapter 05: Flutter App Structure. In P. Tyagi, *Pragmatic Flutter: Building Cross-Platform Mobile Apps for Android, iOS, Web & Desktop*. CRC Press.

Tyagi, P. (2021). Chapter 08: Responsive Interfaces. In P. Tyagi, *Pragmatic Flutter: Building Cross-Platform Mobile Apps for Android, iOS, Web & Desktop*. CRC Press.

8 Responsive Interfaces

This chapter focuses on building responsive layouts for Flutter applications. The responsive layouts can adjust themselves based upon the space available to render their widgets. We will cover `FittedBox` (FittedBox class), `Expanded` (Expanded class), `Flexible` (Flexible class), `FractionallySizedBox` (FractionallySizedBox class), `LayoutBuilder` (LayoutBuilder class), and `Wrap` (Wrap class) widgets.

`FittedBox` WIDGET

The `FittedBox` (FittedBox class) widget fits its child within the given space during layout to avoid overflows. It positions and scales (or clips) its child as per the `fit` property, which helps to fit its child into the space allocated during layout. It makes sure that its child sits inside the parent widget.

Let's understand the usage of `FittedBox` by adding a Row widget with two `Image` widgets as children.

```
Row(
  children: [
    Image.asset('assets/flutter_icon.png'),
    Image.asset('assets/flutter_icon.png'),
  ],
)
```

When the code above is added to the `body` of the `Scaffold` widget, the second image overflows to the right of the screen, as shown in Figure 8.1.

Wrapping the `Row` widget in `FittedBox` makes sure that both of the `Image` widgets are contained inside the `FittedBox` without overflowing out of the screen.

```
FittedBox(
  child: Row(
    children: [
      Image.asset('assets/flutter_icon.png'),
      Image.asset('assets/flutter_icon.png'),
    ],
  ),
)
```

Figure 8.2 shows both `Image` widgets adjusted inside the screen space without overflowing to the right.

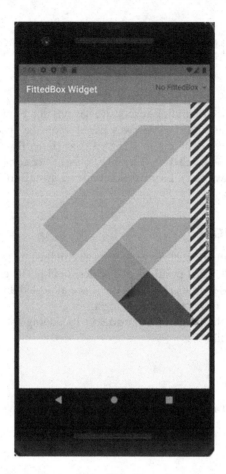

FIGURE 8.1   Overflowing widgets without FittedBox

Source Code Online

The source code for this example (Responsive Interfaces: FittedBox Widget) is available online at GitHub.

Expanded WIDGET

The Expanded (Expanded class) widget is a single-child layout widget. This layout widget is used for a specific child of the multi-layout widgets like Row, Column, and Flex. The child widget wrapped in Expanded widget expands to fill the available space along the main axis. It expands horizontally for Row parent and vertically for Column parent. It uses a flex factor to let the child widget know how much available space they can take up. The child wrapped in the Expanded widget takes up all the open space. It uses the `flex` property to allocate space in case there is a competition between Expanded widgets for available space.

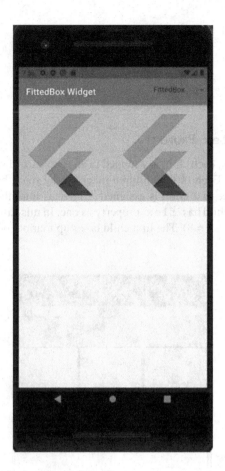

FIGURE 8.2   Responsive layout with FittedBox

Let's check out a widget `expandedDefault()` consisting of Row widget and its three children say `childWidget()`. The `childWidget()` method returns a widget that's added as a child to the Row widget. In the code snippet below, each child is wrapped in an Expanded widget. Each Expanded widget distributes the available horizontal space equally, as shown in Figure 8.3.

```
```
Widget expandedDefault() {
 return Row(
   children: [
     Expanded(
       child: childWidget(""),
     ),
     Expanded(
       child: childWidget(""),
     ),
     Expanded(
```
```

```
        child: childWidget(""),
      ),
    ],
  );
}
```

EXPANDED WITH `flex` PROPERTY

In the example below, each child is wrapped in an `Expanded` widget as well as uses its `flex` property. Each of the children takes up the space based on the value of `flex` property. The first child is assigned `flex` as four, the second is assigned three, and the third child has `flex` property as one. In this case, the total space will be eight parts (4 + 3 + 1 = 8). The first child takes up four out of eight parts or 4/8 of

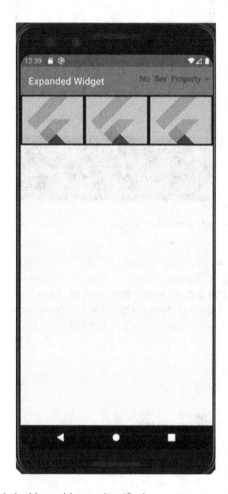

FIGURE 8.3 Expanded widget without using 'flex' property

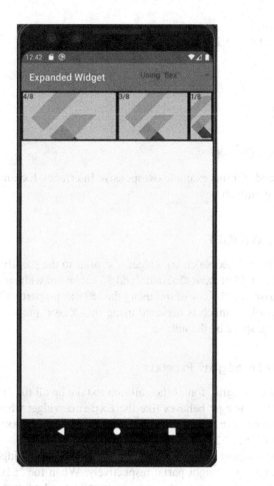

FIGURE 8.4 Expanded widget using 'flex' property

the total space. The second child takes up three out of eight parts or ⅜ of the total space and the last child takes ⅛ of total space, as shown in Figure 8.4.

```
```
Widget expandedWithFlex() {
 return Row(
   children: [
     Expanded(
       flex: 4,
       child: childWidget("4/8"),
     ),
     Expanded(
       flex: 3,
       child: childWidget("3/8"),
     ),
     Expanded(
```
```

```
    flex: 1,
    child: childWidget("1/8"),
  ),
 ],
);
}
```

Source Code Online

The source code for this example (Responsive Interfaces: Expanded Widget) is available online at GitHub.

Flexible WIDGET

The `Flexible` (Flexible class) widget is similar to the `Expand` widget but with a twist. It lets a child of `Row`, `Column`, and `Flex` layout widgets expand in the available space based on the constraint using the `flex` property. Flexible widgets only take space for how much is declared using the `flex` property. They don't claim extra available space by default.

THE `FlexFit.tight` PROPERTY

The `FlexFit.tight` forces the children to take up all the entire available space. The `Flexible` widget behaves like the `Expand` widget when its `fit` property is set to `FlexFit.tight`. The `Expanded(child: Text())` is the same as `Flexible(fit: FlexFit.tight, child: Text())`.

In the code snippet below, the three children of the `Row` widget are assigned four, three, and one out of eight parts, respectively. When the `fit` property is set to `FlexFit.tight`, they take the space assigned to each of them while filling up the available space horizontally.

```
```
Row(
 children: [
   Flexible(
     fit: FlexFit.tight,
     flex: 4,
     child: childWidget("4/8"),
   ),
   Flexible(
     fit: FlexFit.tight,
     flex: 3,
     child: childWidget("3/8"),
   ),
   Flexible(
     fit: FlexFit.tight,
     flex: 1,
```
```

```
      child: childWidget("1/8"),
   ),
 ],
)
```

Refer to Figure 8.5 to see Flexible widget using FlexFit.tight property.

The `FlexFit.loose` Property

The `FlexFit.loose` will keep the default `Flexible` behavior and let them take their maximum sizes. In the code snippet below, the three children to `Row` widget are assigned four, three, and one out of eight parts, respectively, as above. When the `fit` property is set to `FlexFit.loose`, they only take the space assigned to each of them but don't take up the remaining space available horizontally.

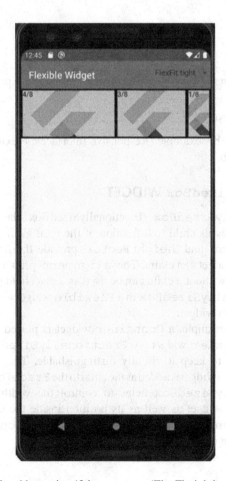

FIGURE 8.5   Flexible widget using 'fit' property as 'FlexFit.tight'

```
` ` `
Row (
 children: [
   Flexible(
     fit: FlexFit.loose,
     flex: 4,
     child: childWidget("4/8"),
   ),
   Flexible(
     fit: FlexFit.loose,
     flex: 3,
     child: childWidget("3/8"),
   ),
   Flexible(
     fit: FlexFit.loose,
     flex: 1,
     child: childWidget("1/8"),
   ),
 ],
)
` ` `
```

Refer to Figure 8.6 to see Flexible widget using FlexFit.loose property.

SOURCE CODE ONLINE

The source code for this example (Responsive Interfaces: Flexible Widget) is available online at GitHub.

FractionallySizedBox WIDGET

The FractionallySizedBox (FractionallySizedBox class) is a single-child layout widget. It sizes its child to a fraction of the total available space. It's properties `widthFactor` and `heightFactor` provide the fraction of the screen real estate that the widget can claim. The `alignment` property positions its child widget. This widget without a child can be used as a placeholder. It's recommended to wrap FractionallySizedBox in a Flexible widget when adding a child to the Row and Column widget.

In the following example, a Container widget is placed in the center of the screen using the Center widget. A FractionallySizedBox is wrapped in a Padding widget to keep it visually distinguishable. The ElevatedButton (ElevatedButton class) widget is added as the child to the FractionallySizedBox. The FractionallySizedBox helps to control the width and height of the ElevatedButton widget as well as its position inside the Container widget using the `alignment` property. In this case, the FractionallySizedBox is aligned to the bottom center to the Container widget.

```
` ` `
Center(
 child: Container(
```

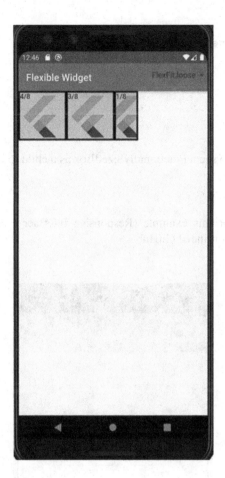

FIGURE 8.6   Flexible widget using 'fit' property as 'FlexFit.loose'

```
width: 200,
height: 200,
decoration: BoxDecoration(
  border: Border.all(),
),
child: Padding(
  padding: const EdgeInsets.all(8.0),
  child: FractionallySizedBox(
    alignment: Alignment.bottomCenter,
    widthFactor: 0.8,
    heightFactor: 0.2,
    child: ElevatedButton(
      child: Text(
        "Tap",
        style: TextStyle(
          fontSize: 20,
        ),
```

```
            ),
          onPressed: () {},
        ),
      ),
    ),
  ),
)
```

Refer to Figure 8.7 to see a FractionallySizedBox as a child to Container widget.

SOURCE CODE ONLINE

The source code for this example (Responsive Interfaces: FractionallySizedBox Widget) is available online at GitHub.

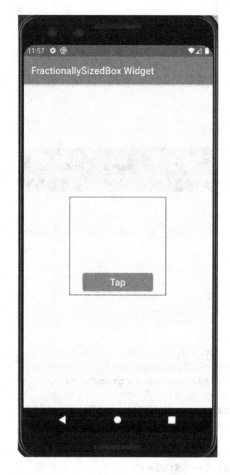

FIGURE 8.7   FractionallySizedBox as a child to Container widget

LayoutBuilder WIDGET

The LayoutBuilder (LayoutBuilder class) widget supports the multi-child layout. It builds widgets dynamically as per the constraint passed by the parent. The LayoutBuilder layout widget works well when creating responsive layouts. It can build appropriate layouts based on the constraints' maximum width (maxWidth) or maximum height (maxHeight). The LayoutBuilder calls the builder function at the layout time.

In this example, LayoutBuilder renders different widgets based on the maximum width of the screen during layout time. It renders `largeScreen()` layout for screens larger than 400 dips whereas `smallScreen()` for any screen with width less than 400 dips.

```
```
LayoutBuilder(
  builder: (context, constraints) {
    if (constraints.maxWidth > 400) {
      return largeScreen();
    } else {
      return smallScreen();
    }
  },
),
```
```

The LayoutBuilder widget rendering on a small screen is shown in Figure 8.8.
The LayoutBuilder widget rendering on a large screen is shown in Figure 8.9.

SOURCE CODE ONLINE

The source code for this example (Responsive Interfaces: LayoutBuilder Widget) is available online at GitHub.

Wrap WIDGET

The Wrap (Wrap class) widget is a layout widget and supports the multi-child layout. This widget is helpful when Row and Column widgets run out of the room. It puts the overflowing widget in the cross-axis when it runs out of space in the placement line along the main-axis. The `direction` property is used to align its children widgets either in the horizontal or vertical direction. The `spacing` property specifies the gap between the children in the same axis. The `runSpacing` property specifies the gap between the runs.

In the following example, the Wrap widget is wrapped in the Center widget. There are six children widgets assigned to the Wrap widget. The child widgets are aligned in the horizontal direction using `direction: Axis.horizontal`. There's a gap of 20 dips in between the children's widgets horizontally. Whenever a number of children overflow the space available in the main axis, the remaining children widgets run in the next line when they're aligned horizontally (or in the

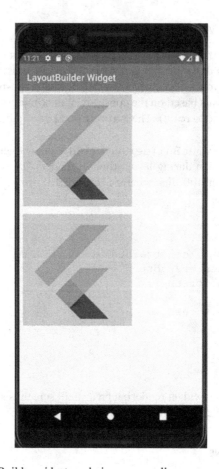

FIGURE 8.8   LayoutBuilder widget rendering on a small screen

next column when aligned vertically). The `runSpacing` property provides the
gap between these runs.

```
```
child: Wrap(
 direction: Axis.horizontal,
 spacing: 20.0,
 runSpacing: 40.0,
 children: [
   childWidget("1"),
   childWidget("2"),
   childWidget("3"),
   childWidget("4"),
   childWidget("5"),
   childWidget("6"),
 ],
),
```
```

FIGURE 8.9   LayoutBuilder widget rendering on a large screen

Refer to Figure 8.10 to see Wrap widget wrapping its children into horizontal/vertical runs.

SOURCE CODE ONLINE

The source code for this example (Responsive Interfaces: Wrap Widget) is available online at GitHub.

CONCLUSION

This chapter concludes our discussion on the Flutter widget. Flutter has a vast library of widgets. This chapter covered some of the frequently used layout widgets to build responsive user interfaces. I encourage you to check out the Flutter widget catalog (Widget Catalog) to learn more about the layout widgets. The

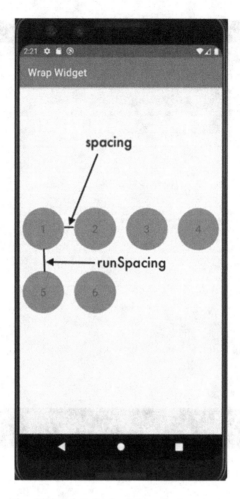

FIGURE 8.10   Wraps the children into horizontal/vertical runs

following concluding points are the general recommendations for building interfaces in Flutter.

- If you know the direction of laying out the widgets, then use Row or Column.
- If you don't know the main axis direction for widgets, then use the Flex widget.
- If you have only one child to display, then use Center or Align to position the child.
- If a child should be smaller than the parent, then wrap it in the Align (Align class) widget.
- If a child is going to be bigger than the parent, then wrap it in a SingleChildScrollView (SingleChildScrollView class) or Overflow (OverflowBox class) widget.

REFERENCES

Flutter Team. (2020, 11 19). *Align class*. Retrieved from Flutter Dev: https://api.flutter.dev/flutter/widgets/Align-class.html

Flutter Team. (2020, 11 19). *ElevatedButton class*. Retrieved from Flutter Dev: https://api.flutter.dev/flutter/material/ElevatedButton-class.html

Flutter Team. (2020, 11 19). *Expanded class*. Retrieved from Flutter Dev: https://api.flutter.dev/flutter/widgets/Expanded-class.html

Flutter Team. (2020, 11 19). *FittedBox class*. Retrieved from Api Flutter Dev: https://api.flutter.dev/flutter/widgets/FittedBox-class.html

Flutter Team. (2020, 11 19). *Flexible class*. Retrieved from Api Flutter Dev: https://api.flutter.dev/flutter/widgets/Flexible-class.html

Flutter Team. (2020, 11 19). *FractionallySizedBox class*. Retrieved from Api Flutter Dev: https://api.flutter.dev/flutter/widgets/FractionallySizedBox-class.html

Flutter Team. (2020, 11 19). *LayoutBuilder class*. Retrieved from Api Flutter Dev: https://api.flutter.dev/flutter/widgets/LayoutBuilder-class.html

Flutter Team. (2020, 11 19). *OverflowBox class*. Retrieved from Flutter Dev: https://api.flutter.dev/flutter/widgets/OverflowBox-class.html

Flutter Team. (2020, 11 19). *SingleChildScrollView class*. Retrieved from Flutter Dev: https://api.flutter.dev/flutter/widgets/SingleChildScrollView-class.html

Flutter Team. (2020, 11 19). *Widget Catalog*. Retrieved from Flutter Api Dev: https://flutter.dev/docs/development/ui/widgets

Flutter Team. (2020, 11 19). *Wrap class*. Retrieved from Api Flutter Dev: https://api.flutter.dev/flutter/widgets/Wrap-class.html

Tyagi, P. (2020, 11 19). *Responsive Interfaces: Expanded Widget*. Retrieved from Chapter08: Pragmatic Flutter GitHub Repo: https://github.com/ptyagicodecamp/pragmatic_flutter/blob/master/lib/chapter08/layouts/expanded.dart#L84

Tyagi, P. (2020, 11 19). *Responsive Interfaces: FittedBox Widget*. Retrieved from Chapter08: Pragmatic Flutter GitHub Repo: https://github.com/ptyagicodecamp/pragmatic_flutter/blob/master/lib/chapter08/layouts/fitted_box.dart#L80:L84

Tyagi, P. (2020, 11 19). *Responsive Interfaces: Flexible Widget*. Retrieved from Chapter08: Pragmatic Flutter GitHub Repo: https://github.com/ptyagicodecamp/pragmatic_flutter/blob/master/lib/chapter08/layouts/flexible.dart#L77:L96

Tyagi, P. (2020, 11 19). *Responsive Interfaces: FractionallySizedBox Widget*. Retrieved from Chapter08: Pragmatic Flutter GitHub Repo: https://github.com/ptyagicodecamp/pragmatic_flutter/blob/master/lib/chapter08/layouts/fractionally_sized_box.dart#L27:L49

Tyagi, P. (2020, 11 19). *Responsive Interfaces: LayoutBuilder Widget*. Retrieved from Chapter08: Pragmatic Flutter GitHub Repo: https://github.com/ptyagicodecamp/pragmatic_flutter/blob/master/lib/chapter08/layouts/layoutbuilder.dart

Tyagi, P. (2020, 11 19). *Responsive Interfaces: Wrap Widget*. Retrieved from Chapter08: Pragmatic Flutter GitHub Repo: https://github.com/ptyagicodecamp/pragmatic_flutter/blob/master/lib/chapter08/layouts/wrap.dart#L27:L39

9 Building User Interface for BooksApp

In this chapter, we'll take the *HelloBooksApp* created earlier in this book (Chapter 05: Flutter App Structure) to the next level. Let's rename it to *BooksApp*. The *BooksApp* lists books' titles and their authors in a list view.

THE *BooksApp* INTERFACE

The book listing in the *BooksApp* will look like below for each of the four platforms.

ANDROID

BooksApp's primary user interface for the Android platform (Figure 9.1).

iOS

BooksApp's primary user interface for the iOS platform (Figure 9.2).

WEB

BooksApp's primary user interface for the web platform (Figure 9.3).

DESKTOP (macOS)

BooksApp's primary user interface for the desktop-macos platform (Figure 9.4).

THE *BooksApp* ANATOMY

Let's take a look at the *BooksApp*'s widget structure. The `MaterialApp` widget is at the root of the *BooksApp*. It has a `Scaffold` widget as its child. The `Scaffold` widget has an `AppBar` widget and `ListView` widget for its `body` property. The `ListView.builder()` is used to build the `ListView` of `Card` (Card class) widgets. Each `Card` widget displays title, author, and image information for the book. We're using mocked sample book data for demonstration purposes. It has two book entries. The `ListView` has two children of `Card` widgets, as shown in the app structure diagram (Figure 9.5).

The `Card` widget displays title, author list, and cover image for the book. The `Card` widget has a `Padding` widget as its child. The `Padding` widget insets its child `Row` widget. The padding makes sure that the content of the `Row` widget is not bleeding over the edges. The `Row` widget displays its children horizontally. The `Row`

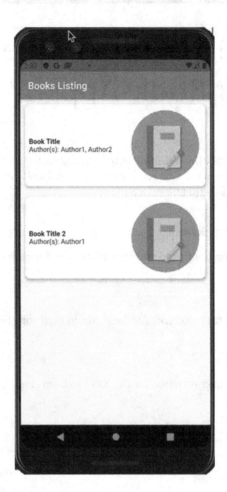

FIGURE 9.1   BooksApp – Android

widget has `Flexible` and `Image` as its children. The `Flexible` widget displays the book's title and author list with the title at the top and author list to the bottom. The `Column` widget is appropriate to align its children in vertical alignment. The `Column` widget has two `Text` widgets as its children. The first `Text` widget is to display the title of the book, and the second `Text` widget is to display the author list (Figure 9.6).

IMPLEMENTING USER INTERFACE

The `BooksApp` extends `StatelessWidget`. The book listing is constructed in its own `BooksListing` widget. The `BooksListing` widget is a `StatelessWidget` as well. The book listing data is hardcoded in the `bookData()` method for demonstration purposes. Refer to Figure 9.5 for visual understanding of the widgets discussed in this section.

FIGURE 9.2   BooksApp – iOS

BOOKS SAMPLE JSON DATA

The books data is being hardcoded and provided using the `bookData()` function as below:

```
List bookData() {
    return [
        {'title':'Book Title','authors':['Author1',
    'Author2'], 'image':'assets/book_cover.png'},
        {'title':'Book Title 2','authors':['Author1'],
    'image':'assets/book_cover.png'}
    ];
}
```

The first book entry has two authors and the second book has only one author. The same cover art is used for both books to keep it simple. The `bookData()` function returns a List of JSON (Introducing JSON) entries for each book. The JSON stands for 'JavaScript Object Notation'. It is data interchange format. This data format is

FIGURE 9.3   BooksApp – Web

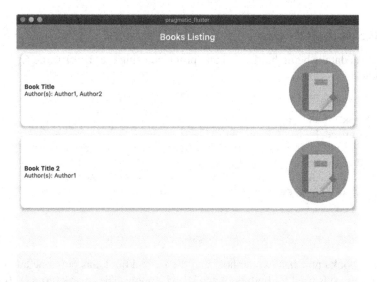

FIGURE 9.4   BooksApp –Desktop (macos)

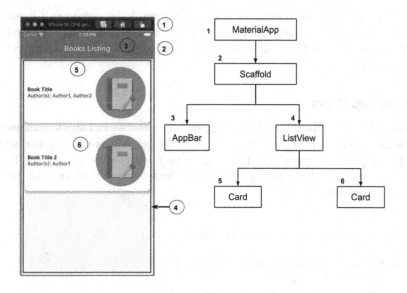

FIGURE 9.5 Anatomy of BooksApp

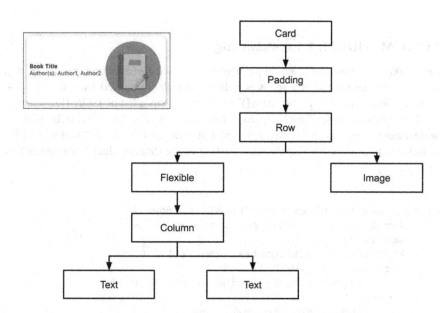

FIGURE 9.6 Anatomy of BooksApp's card widget

used for transferring data from one source to another. In later chapters (Chapter 12: Integrating REST API) and (Chapter 13: Data Modeling), you'll learn to fetch book information from JSON formatted data entries and build an interface to display book information.

BooksApp WIDGET

The BooksApp is the StatelessWidget. It uses MaterialApp in combination with the Scaffold widget. The Scaffold widget assigns the AppBar widget to the `appBar` property. The BooksListing widget is assigned to its `body` property. The `booksListing` is populated from data returned from the `book-Data()` function.

```
class BooksApp extends StatelessWidget {
  @override
  Widget build(BuildContext context) {
    return MaterialApp(
      debugShowCheckedModeBanner: false,
      home: Scaffold(
        appBar: AppBar(
          title: Text("Books Listing")),
        body: BooksListing(),
      ),
    );
  }
}
```

CUSTOM WIDGET: BooksListing

Let's take a deeper look into implementing the BooksListing widget in this section. The `booksListing` is populated with data returned from the `book-Data()` function. The `bookData()` function returns a list of JSON (Introducing JSON) entries for books. One JSON entry for each book. The `ListView.builder` widget iterates over each item of `booksListing`, and creates a Card widget for it. Refer to Figure 9.6 for visual understanding of the Card widget discussed in this section.

```
class BooksListing extends StatelessWidget {
  final booksListing = bookData();
  @override
  Widget build(BuildContext context) {
    return ListView.builder(
      itemCount: booksListing == null?  0 :
    booksListing.length,
      itemBuilder: (context, index) {
        return Card();
```

```
        }
}
```

A shape border can be drawn around the `Card` (Card class) widget. We are creating a rectangle border with rounded corners. The corner's radius is assigned using the `borderRadius` property. The `Card` widget can be given elevation using the `elevation` property. The `margin` property is used to provide card spacing around it.

```
Card(
    shape: RoundedRectangleBorder(
        borderRadius: BorderRadius.circular(10.0),
    ),
    elevation: 5,
    margin: EdgeInsets.all(10),
);
```

Padding WIDGET

The `Padding` (Padding class) widget is added as a child to the `Card` widget, as shown in Figure 9.6. It provides the inset around its child. We'll give a default padding of 8 logical pixels.

```
Card(
  child: Padding(
    padding: const EdgeInsets.all(8.0),
  ),
);
```

Row WIDGET

The `Row` widget displays its children in a horizontal array. We'll use this array to display the book's title, authors' list to the left, and cover image to the screen's right. The `Row` widget's `mainAxisAlignment` attribute aligns the children widgets along its main axis, which is horizontal in this case. The `MainAxisAlignment.spaceBetween` property distributes free space available evenly between the children widgets. The first child is the `Flexible` widget.

```
Card(
  child: Padding(
    child: Row(
      mainAxisAlignment: MainAxisAlignment.spaceBetween,
      children: [
```

```
     Flexible(),
     Image.asset(
       booksListing[index]['image'],
       fit: BoxFit.fill,
     ),
   ],
 ),
),
);
```

Flexible WIDGET

The Flexible widget renders the book's title and its authors' list in a Column widget. A Column widget renders its children vertically. The Flexible widget gives flexibility to its child, the Column widget's children flexibility to expand to fill the available space in its main axis. The main axis for the Column widget is vertical. The book's title and authors' list expand vertically rather than horizontally and hereby giving ample space for the cover image to render to the right side of the screen.

```
Flexible(
 child: Column(
   crossAxisAlignment: CrossAxisAlignment.start,
   children: <Widget>[
     Text(
       '${booksListing[index]['title']}',
       style: TextStyle(fontSize: 14, fontWeight:
FontWeight.bold),
     ),
     booksListing[index]['authors'] != null
        ? Text(
          'Author(s): ${booksListing[index]['authors'].join(",
")}',
          style: TextStyle(fontSize: 14),
        )
          : Text(""),
   ],
 ),
),
```

In the code above, the `booksListing[index]['title']` provides the title of the book. The TextStyle provides the styling like font size and font weight for the title text. The `booksListing[index]['authors']` provides the authors' list. The list items are joined with a comma and display the items in the list as one string altogether. However, there could be cases when there's no explicit authors' list available for a given book. It's a good practice to check for the data availability and show the widgets accordingly. If the authors' list is empty, then show a Text widget with an empty string.

Image WIDGET

The cover art for the book is displayed using the Image widget. In this app, the book cover art is available in the 'assets' folder of the Flutter project root. In this example, the sample book cover is the same for both images and is named 'book_cover.png'. Don't forget to add the images under the 'assets' section of the 'pubspec.yaml' configuration file.

```
assets:
  - assets/book_cover.png
```

The image path is retrieved as `booksListing[index]['image']`. If the path is not available, then show an empty placeholder Container widget. The `BoxFit. fill` property is used to fill the image in the target box.

```
booksListing[index]['image'] != null
    ? Image.asset(
  booksListing[index]['image'],
  fit: BoxFit.fill,
)
    : Container(),
```

SOURCE CODE ONLINE

The code for this example (Building User Interface for BooksApp) is available online at GitHub.

CONCLUSION

In this chapter, you learned to create the layout for the user interface for *BooksApp*. Flutter widgets introduced in previous chapters like Column, Row, Padding, Flexible, Image, ListView are used to implement the interface. The Card widget is used to display the book's title and authors' list. You briefly touched on to parse book information from JSON formatted data entries and build an interface to display book information.

REFERENCES

Flutter Dev. (2020, 12 21). *Card class*. Retrieved from Flutter Dev: https://api.flutter.dev/flutter/material/Card-class.html
Google. (2020, 11 20). *Card class*. Retrieved from Flutter Dev: https://api.flutter.dev/flutter/material/Card-class.html

Google. (2020, 11 20). *Padding class*. Retrieved from Api Flutter Dev: https://api.flutter.dev/
flutter/widgets/Padding-class.html

json.org. (2020, 11 20). *Introducing JSON*. Retrieved from json.org: https://www.json.org/

Tyagi, P. (2020, 11 20). *Building User Interface for BooksApp*. Retrieved from Chapter09:
Pragmatic Flutter GitHub Repo: https://github.com/ptyagicodecamp/pragmatic_flutter/
blob/master/lib/chapter09/main_09.dart

Tyagi, P. (2021). Chapter 05: Flutter App Structure. In P. Tyagi, *Pragmatic Flutter: Building
Cross-Platform Mobile Apps for Android, iOS, Web & Desktop*. CRC Press.

Tyagi, P. (2021). Chapter 12: Integrating REST API. In P. Tyagi, *Pragmatic Flutter: Building
Cross-Platform Mobile Apps for Android, iOS, Web & Desktop*. CRC Press.

Tyagi, P. (2021). Chapter 13: Data Modeling. In P. Tyagi, *Pragmatic Flutter: Building Cross-
Platform Mobile Apps for Android, iOS, Web & Desktop*. CRC Press.

10 Flutter Themes

In this chapter, you will learn the basics of Flutter themes and how to use it to style your apps. Flutter theme is used to define application-wide stylings like colors, fonts, and text. There are two ways to implement themes in Flutter:

- Global Theme: Styling throughout the application. This chapter will define the global theme to style the *BooksApp* developed in the previous chapter (Chapter 09: Building User Interface for BooksApp).
- Local Theme: Styling a specific widget. We will use the local theme to style one part of the *BooksApp*. We will style the `BooksListing` widget displaying books' information without impacting other parts of the app.

GLOBAL THEME

A global theme is used to style the entire app all at once. Global themes are implemented using ThemeData (ThemeData class). We will create and use a Global theme for *BooksApp* using ThemeData to hold styling information for the material apps. The ThemeData widget uses the following properties to define styling attributes.

- The `brightness` (brightness property) property uses the `brightness` property to assign light or dark color themes to the app.
- The `appBarTheme` (appBarTheme property) property: The `appBarTheme` property defines the theme for the AppBar widget.
- The `iconTheme` (iconTheme property) property: The `iconTheme` property of AppBarTheme declares theme selection for the icons used in the app bar. However, the iconTheme property for the ThemeData widget defines the icon colors for the whole app globally. The IconThemeData (IconThemeData class) class defines the color, size, and opacity of icons.

A 'Home' icon is added to the AppBar (AppBar class) widget to demonstrate styling icons in an app. Let's create two light themes and one dark theme to understand creating and using themes.

DEFAULT BLUE THEME

The blue theme comes by default with the Flutter starter application. The AppBar's color scheme can be customized using the `appBarTheme` property. The default light blue theme is stored in the `defaultTheme` variable.

```
ThemeData defaultTheme = ThemeData(
```

```
// Define the default brightness and colors for the overall
app.
brightness: Brightness.light,
primaryColor: Colors.blue,
accentColor: Colors.lightBlueAccent,
appBarTheme: AppBarTheme(
  color: Colors.blue,
  iconTheme: IconThemeData(
    color: Colors.white,
  ),
),
);
```

USING THE DEFAULT THEME

The `defaultTheme` is assigned to the `theme` property of MaterialApp. The *BooksApp* in default theme is shown in Figure 10.1.

FIGURE 10.1 Default light global theme

```
~ ~ ~
MaterialApp(
  theme: defaultTheme,
  home: Scaffold(
    appBar: AppBar(
      leading: Icon(Icons.home),
      title: Text("Books Listing"),
    ),
    body: BooksListing(),
  ),
);
~ ~ ~
```

LIGHT PINK THEME

Let's change the primary and accent color to pink. Unless there's an app bar specific styling needed, you don't need to specify the `appBarTheme` property. The pink theme is stored in the `pinkTheme` variable.

```
~ ~ ~
ThemeData pinkTheme = ThemeData(
  // Define the default brightness and colors for the overall
app.
  brightness: Brightness.light,
  primaryColor: Colors.pink,
  accentColor: Colors.pinkAccent,
);
~ ~ ~
```

USING PINK THEME

The `pinkTheme` is assigned to the `theme` property of MaterialApp. The *BooksApp* with pink colored theme is shown in Figure 10.2.

```
~ ~ ~
MaterialApp(
  theme: pinkTheme,
  home: Scaffold(
    appBar: AppBar(
      leading: Icon(Icons.home),
      title: Text("Books Listing"),
    ),
    body: BooksListing(),
  ),
);
~ ~ ~
```

DARK THEME

Let's define a dark theme for the *BooksApp*. The brightness property is set to dark. Primary and accent colors are modified to colors appropriate for a darker

FIGURE 10.2   Pink light theme

theme. Later in the chapter, you will learn to switch from one theme to another. The dark theme is stored in the `darkTheme` variable.

```
```
ThemeData darkTheme = ThemeData(
 // Define the default brightness and colors for the overall app.
 brightness: Brightness.dark,
 primaryColor: Colors.orange,
 accentColor: Colors.yellowAccent,
);
```
```

USING DARK THEME

The `darkTheme` is assigned to the `theme` property of `MaterialApp`. The BooksApp with dark theme is shown in Figure 10.3.

```
```
MaterialApp(
 theme: darkTheme,
```
```

FIGURE 10.3 Dark theme

```
home: Scaffold(
  appBar: AppBar(
    leading: Icon(Icons.home),
    title: Text("Books Listing"),
  ),
  body: BooksListing(),
),
);
~ ~ ~
```

SOURCE CODE ONLINE

Source code for this example (Flutter Themes) is available at GitHub.

MODULARIZING THEMES

All themes can be kept in a separate file. This file needs to be imported into the referencing file. This helps to keep all the styling code together. You may be tempted to create a class with one static method for each theme. However, a class with only static methods is discouraged per this lint rule (avoid_classes_with_only_static_members).

If utility methods are not logically related in Dart, they shouldn't be put inside a class. Such methods can be put at top-level in a dart file.

All global themes can be put in a file *themes.dart* to keep all themes together.

FILE: THEMES.DART

```
``` 

import 'package:flutter/material.dart';

ThemeData get defaultTheme => ThemeData(
    // Define the default brightness and colors for the
overall app.
    brightness: Brightness.light,
    primaryColor: Colors.blue,
    accentColor: Colors.lightBlueAccent,
    appBarTheme: AppBarTheme(
      color: Colors.blue,
      iconTheme: IconThemeData(
        color: Colors.white,
      ),
    ),
  );

ThemeData get pinkTheme => ThemeData(
    // Define the default brightness and colors for the
overall app.
    brightness: Brightness.light,
    primaryColor: Colors.pink,
    accentColor: Colors.pinkAccent,
  );

ThemeData get darkTheme => ThemeData(
    // Define the default brightness and colors for the
overall app.
    brightness: Brightness.dark,
    primaryColor: Colors.orange,
    accentColor: Colors.yellowAccent,
  );
```
```

IMPORTING THEMES

The '*themes.dart*' is imported to access the `defaultTheme` from MaterialApp.

```
```

import 'themes.dart';
MaterialApp(
  theme: defaultTheme,
  home: Scaffold(
    appBar: AppBar(
      leading: Icon(Icons.home),
      title: Text("Books Listing"),
```
```

```
    ),
    body: BooksListing(),
  ),
);
```

SOURCE CODE ONLINE

Source code for this example (Flutter Themes: Modularizing Themes) is available at GitHub.

USING CUSTOM FONTS

DOWNLOAD FONT

The first step is to download the font that you want to use. I have downloaded the Pangolin (Pangolin) font from Google Fonts. Copy the '*.ttf' file into the Flutter root project's *assets* directory. I have created a '*font*' directory under '*assets*' to keep the fonts-related files in one place.

CONFIGURATION

Once you have got a TTF file copied into the Flutter project, it's time to add it in the '*pubspec.yaml*' configuration.

```
fonts:
  - family: Pangolin
    fonts:
      - asset: assets/fonts/Pangolin-Regular.ttf
```

USING CUSTOM FONT

Let's see how to use this font to style the AppBar's title (Figure 10.4).

```
MaterialApp(
  theme: defaultTheme,
  home: Scaffold(
    appBar: AppBar(
      leading: Icon(Icons.home),
      title: Text(
        "Books Listing",
        style: TextStyle(fontFamily: 'Pangolin', fontSize: 30),
      ),
    ),
    body: BooksListing(),
  ),
);
```

FIGURE 10.4   Custom font in AppBar title

SOURCE CODE ONLINE

Source code for this example (Flutter Themes: Using Custom Fonts) is available at
GitHub.

LOCAL THEME

The local theme is used to customize the theme of a part of the screen rather than the
whole app. The local theme is applied by wrapping the target widget in the `Theme`
widget and providing the custom `ThemeData` for its `data` property like below:

```
Theme (
    data: ThemeData (
      //Implement custom theme here
      ),
      //Theme is applied to TargetWidget
      child: TargetWidget,
);
```

In the *BooksApp*, we'll learn to create and use a local theme for making `Card` widget's color pink, customizing text themes for the book title, and authors' listing widgets.

Card COLOR

The `Card` widget is wrapped as a child to the `Theme` (Theme class) widget. This widget applies the theme to its descendant widget(s) by assigning `ThemeData` to the `data` property of the `Theme` widget, as shown in the code snippet below. The `Card` widget is assigned a pink color using the `cardColor` property of `ThemeData`. Refer to Figure 10.5 to observe the pink color for the `Card` widget.

```
```
Theme (
 //ThemeData local to Card widget
 data: ThemeData(
    cardColor: Colors.pinkAccent,
 ),
 child: ListView.builder(
    itemCount: booksListing == null? 0 : booksListing.length,
    itemBuilder: (context, index) {
```

**FIGURE 10.5** Local theme – Card color, book title TextTheme, extending parent theme for book's authors' list TextTheme
```

```
   return Card(
     ...
     );
   },
 ),
);
```

Book's Title `TextTheme`

We'll use the custom font 'Pangolin' to style the book's title in the `Card` widget. We'll create a local text theme for the book's title. The `TextTheme` (TextTheme class) widget is used to create a text theme. It applies material design text themes. Its `headline6` property is assigned to a custom style. The book's title applies this custom text theme by assigning its `style` property to `headline6`. The `headline6` style is accessed using `Theme.of(context).textTheme.headline6`.

```
Theme(
 data: ThemeData(
   ...,
   textTheme: TextTheme(
     headline6: TextStyle(fontFamily: 'Pangolin', fontSize:
20),
   ),
 ),
 child: ListView.builder(
   itemCount: booksListing == null?  0 : booksListing.length,
   itemBuilder: (context, index) {
     return Card(
       ...
         Text(
         '${booksListing[index]['title']}',
         //Using custom text theme
         style: Theme.of(context).textTheme.headline6,
         ),
         ...
     );
   },
 ),
);
```

Extending Parent's `TextTheme`

The parent's theme can be extended to customize to your needs using the `copyWith` (copyWith method) method. We will customize the parent's text theme `bodyText2` to the italic font style. The `bodyText2` is accessed from book's authors' list `Text` widget's style using `Theme.of(context).textTheme.bodyText2`.

```
` ` `
Theme(
 data: ThemeData(
   ...,
   textTheme: TextTheme(
         bodyText2: Theme.of(context)
         .copyWith(
           textTheme: TextTheme(
           bodyText2: TextStyle(fontStyle: FontStyle.italic),
           ),
         ). textTheme.bodyText2,
    ),
 ),
 child: ListView.builder(
   itemCount: booksListing == null?  0 : booksListing.length,
   itemBuilder: (context, index) {
     return Card(
         ...
         Text(
         'Author(s): ${booksListing[index]['authors'].join(",
")}',
         //Using custom text theme
         style: Theme.of(context).textTheme.bodyText2,
         ),
         ...
     );
   },
 ),
);
` ` `
```

SOURCE CODE ONLINE

Source code of this example (Flutter Themes: Local Theme) is available at GitHub.

SWITCHING THEMES

So far, you have learned to create different types of themes. In real-world applications, you may want to provide functionality in your app to switch from one type of theme to another. It's a popular use case to toggle between light to dark themes and vice versa. For this purpose, the app needs to remember the state of the app. The *BooksApp* widget needs to be a StatefulWidget to keep track of the currently selected theme. The themes are represented using enumeration `AppThemes`.

```
` ` `
enum AppThemes { light, dark }
` ` `
```

The ` _ BooksAppState` holds the `currentTheme`.

```
` ` `
class _BooksAppState extends State<BooksApp> {
 var currentTheme = AppThemes.light;
}
` ` `
```

The `MaterialApp` use `theme` property to apply the currently selected theme as below:

```
` ` `
MaterialApp(
   ...
     theme: currentTheme == AppThemes.light?  defaultTheme :
darkTheme,
     ...
}
` ` `
```

Note: The `defaultTheme` and `darkTheme` are defined in *themes.dart* file, and are imported using `import 'themes.dart'`.

Switching Themes

An `IconButton` (IconButton class) widget is added in the `AppBar` widget to facilitate theme switching. Pressing on this icon button toggles the `currentTheme`. If a light theme is selected, then `currentTheme` is assigned to `darkTheme` or vice versa. The `currentTheme` is updated inside the `setState` method. The updated value forces `MaterialApp` to rebuild and applies the currently selected theme.

```
` ` `
appBar: AppBar(
   ...
    actions: [
      IconButton(
        icon: Icon(Icons.all_inclusive),
        // Toggling from light to dark theme and vice versa
        onPressed: () {
          setState(() {
            currentTheme = currentTheme == AppThemes.light
                ? AppThemes.dark
                : AppThemes.light;
          });
        },
      )
    ])
` ` `
```

The *BooksApp* widget look like as below:

```
` ` `
import 'themes.dart';
```

```
//BooksApp's entry point
void main() => runApp(BooksApp());

//StatefulWidget
class BooksApp extends StatefulWidget {
 @override
 _BooksAppState createState() => _BooksAppState();
}
class _BooksAppState extends State<BooksApp> {
 var currentTheme = AppThemes.light;
 @override
 Widget build(BuildContext context) {
   return MaterialApp(
     debugShowCheckedModeBanner: false,
     //NEW CODE: applying selected theme
     theme: currentTheme == AppThemes.light?  defaultTheme :
darkTheme,
     home: Scaffold(
       appBar: AppBar(
           leading: Icon(Icons.home),
           title: Text("Books Listing"),
           actions: [
             IconButton(
               icon: Icon(Icons.all_inclusive),
               //NEW CODE: Toggling from light to dark theme
and vice versa
               onPressed: () {
                 setState(() {
                   currentTheme = currentTheme ==
AppThemes.light
                     ? AppThemes.dark
                     : AppThemes.light;
                 });
               },
             )
           ]),
       body: BooksListing(),
     ),
   );
 }
}
```

The `BookListing` widget remains the `StatelessWidget` because there's no need to update this part of the interface.

LIGHT THEME

The default selected theme is a light blue theme, as shown in the screenshot (Figure 10.6). The default theme is defined as below in the '*themes.dart*' file:

FIGURE 10.6   Default light global theme for BooksApp

```
```
ThemeData(
    // Define the default brightness and colors for the
overall app.
    brightness: Brightness.light,
    primaryColor: Colors.blue,
    accentColor: Colors.lightBlueAccent,
    appBarTheme: AppBarTheme(
      color: Colors.blue,
      iconTheme: IconThemeData(
        color: Colors.white,
      ),
    ),
  );
```
```

DARK THEME

The dark theme is shown in the screenshot (Figure 10.7). The ThemeData for the dark theme is defined as below in the `themes.dart` file:

FIGURE 10.7   After switching to dark global theme for BooksApp

```
~ ~ ~
ThemeData(
    // Define the default brightness and colors for the
overall app.
    brightness: Brightness.dark,
    primaryColor: Colors.orange,
    accentColor: Colors.yellowAccent,
    );
~ ~ ~
```

SOURCE CODE ONLINE

The full source code of this example (Flutter Themes: Switching Themes) is available at GitHub.

CONCLUSION

In this chapter, you learned the basics of creating and using custom themes. The global and local themes were created for the *BooksApp* application. The custom fonts were used to style the book's title text in the `Card` widget. Two global themes, light and dark, were created for the app. Lastly, you also learned to toggle from one theme to another.

REFERENCES

Burke, K. (2020, 11 20). *Pangolin*. Retrieved from Google Fonts: https://fonts.google.com/specimen/Pangolin#standard-styles

Dart Team. (2020, 11 20). *avoid_classes_with_only_static_members*. Retrieved from Dart language: https://dart-lang.github.io/linter/lints/avoid_classes_with_only_static_members.html

Google. (2020, 11 20). *AppBar class*. Retrieved from Flutter Dev: https://api.flutter.dev/flutter/material/AppBar-class.html

Google. (2020, 11 20). *appBarTheme property*. Retrieved from Flutter Dev: https://api.flutter.dev/flutter/material/ThemeData/appBarTheme.html

Google. (2020, 11 20). *brightness property*. Retrieved from Flutter Dev: https://api.flutter.dev/flutter/material/AppBar/brightness.html

Google. (2020, 11 20). *copyWith method*. Retrieved from Flutter Dev: https://api.flutter.dev/flutter/material/TextTheme/copyWith.html

Google. (2020, 11 20). *IconButton class*. Retrieved from Flutter Dev: https://api.flutter.dev/flutter/material/IconButton-class.html

Google. (2020, 11 20). *iconTheme property*. Retrieved from Flutter Dev: https://api.flutter.dev/flutter/material/AppBar/iconTheme.html

Google. (2020, 11 20). *IconThemeData class*. Retrieved from Flutter Dev: https://api.flutter.dev/flutter/widgets/IconThemeData-class.html

Google. (2020, 11 20). *TextTheme class*. Retrieved from Flutter Dev: https://api.flutter.dev/flutter/material/TextTheme-class.html

Google. (2020, 11 20). *Theme class*. Retrieved from Flutter Dev: https://api.flutter.dev/flutter/material/Theme-class.html

Google. (2020, 11 20). *ThemeData class*. Retrieved from Flutter Dev: https://api.flutter.dev/flutter/material/ThemeData-class.html

Tyagi, P. (2020, 11 20). *Flutter Themes*. Retrieved from Chapter10: Pragmatic Flutter GitHub Repo: https://github.com/ptyagicodecamp/pragmatic_flutter/blob/master/lib/chapter10/main_10_global.dart

Tyagi, P. (2020, 11 20). *Flutter Themes: Local Theme*. Retrieved from Chapter10: Pragmatic Flutter GitHub Repo: https://github.com/ptyagicodecamp/pragmatic_flutter/blob/master/lib/chapter10/main_10_local.dart

Tyagi, P. (2020, 11 20). *Flutter Themes: Modularizing Themes*. Retrieved from Chapter10: Pragmatic Flutter GitHub Repo: https://github.com/ptyagicodecamp/pragmatic_flutter/blob/master/lib/chapter10/main_10_global_modularize.dart

Tyagi, P. (2020, 11 20). *Flutter Themes: Switching Themes*. Retrieved from Chapter10: Pragmatic Flutter GitHub Repo: https://github.com/ptyagicodecamp/pragmatic_flutter/blob/master/lib/chapter10/main_10_switchingThemes.dart

Tyagi, P. (2020, 11 20). *Flutter Themes: Using Custom Fonts*. Retrieved from Chapter10: Pragmatic Flutter GitHub Repo: https://github.com/ptyagicodecamp/pragmatic_flutter/blob/master/lib/chapter10/main_10_global_customfont.dart

Tyagi, P. (2021). Chapter 09: Building User Interface for BooksApp. In P. Tyagi, *Pragmatic Flutter: Building Cross-Platform Mobile Apps for Android, iOS, Web & Desktop*. CRC Press.

11 Persisting Data

This chapter introduces persisting data in a Flutter application. There are two approaches to persist data on the disk permanently in Flutter applications. In the previous chapter (Chapter 10: Flutter Themes, 2021) you learned to toggle application's themes from one to another. However, that change doesn't persist from one launch to another. Users have to select a theme every time they launch the app. This is not a very good user experience. It makes sense to persist the theme's selection from the previous launch, so that users can have seamless experience across the multiple launches. In this chapter, you will learn to persist theme selection using two approaches: Key/Value datastore (Shared preferences) and Local database. Shared preferences are the better choice when there's a tiny amount of data that needs to be stored. Local database is a better choice for a huge dataset. In real-world applications, it makes sense to use the Shared preference approach to save theme selection. The difference between persisting data using Shared preferences plugin vs Local database is that Shared preferences plugin cannot guarantee persisting writes to disk after app restarts. However, saving data to a Local database is more reliable. In this chapter, both approaches are demonstrated to persist data for the selected theme on Android, iOS, web, and desktop-macOS platforms.

LIGHT THEME (DEFAULT)

The *BooksApp* launches for the first time in the default blue theme. The *BooksApp*'s theme can be changed to the dark theme by clicking on the app bar's infinity icon. However, if you exit the app and restart it again it restores to the light blue theme. This happened because the theme selection was not persisted permanently (Figure 11.1).

DARK THEME

The *BooksApp*'s dark theme is shown in Figure 11.2. The app's theme can be switched back to light theme by clicking on the same app bar's infinity icon. However, none of the theme selection is persisted to the disk and will be restored to the default theme every time the app restarts.

In the following sections, you will learn to save the selected theme preference to the disk using key/value datastore and Local database variations.

KEY/VALUE DATA STORE (SHARED PREFERENCES PLUGIN)

The Shared preference approach uses key/value pairs to store information on the device using Shared Preference Flutter plugin (Shared preferences plugin). Shared preference plugin is useful in storing small sized data using key-value format on disk. This solution is useful when amount of data to be saved is relatively small like saving user preferences. In the previous chapter, you learned how to implement

FIGURE 11.1   Persisted light theme applied to BooksApp

theme switching from default light theme to dark theme and vice versa, by clicking on `AppBar`'s infinity icon. In this section, we'll see how to persist the selected
theme using the Shared preferences plugin. The iOS platform uses NSUserDefaults
(NSUserDefaults) and Android platform uses SharedPreferences (SharedPreferences)
implementations to store simple data as key-value pairs to disk asynchronously.

The *pubspec.yaml* Dependency

The first step is to add the Shared preference plugin to the project's dependencies in
the *pubspec.yaml* file. At the time of this writing the *shared_preferences* plugin's
version is 0.5.8. Please update it to the latest version.

```
```
dependencies:
 flutter:
   sdk: flutter
 #SharedPreference-persisting theme selection
 shared_preferences: ^0.5.8
```
```

FIGURE 11.2   Persisted dark theme applied to BooksApp

LOADING THEME

The ` _ BooksAppState` class holds the active theme in the `currentTheme` variable. The default theme is `AppThemes.light`. The theme settings are restored from persistent store during the app's startup from the `initState()` method. The `loadActiveTheme(.)` gets the reference to Shared preference using `SharedPreferences.getInstance()` which is an asynchronous operation. The `await/async` keywords are used to get reference to `SharedPreferences` class. The `AppThemes` enumeration's index is used as theme id, and saved in the Shared preference using key `theme _ id`. The app's first launch will return `null` for `sharedPrefs.getInt('theme _ id')`, and `AppThemes. light.index` is assigned as default `themeId`. Once `themeId` is available, the `currentTheme` is assigned to the selected theme using `AppThemes. values[themeId]`. The `currentTheme` is called from the `setState()` method that rebuilds the `MaterialApp` widget, and hence reassigns `current-Theme` to the `theme` property of the `MaterialApp`.

```
` ` `
import 'package:shared_preferences/shared_preferences.dart';
...
class _BooksAppState extends State<BooksApp> {
 AppThemes currentTheme = AppThemes.light;

 // Fetching theme_id from SharedPreference
 void loadActiveTheme(BuildContext context) async {
   var sharedPrefs = await SharedPreferences.getInstance();
   //if theme_id key is null (not found), then set default
theme
   int themeId = sharedPrefs.getInt('theme_id')? ?
AppThemes.light.index;

   setState(() {
     currentTheme = AppThemes.values[themeId];
   });
 }

 @override
 void initState() {
   super.initState();
   // Load theme from sharedPreference
   loadActiveTheme(context);
 }

 @override
 Widget build(BuildContext context) {
   return MaterialApp(
     //Applying theme to the app
     theme: currentTheme == AppThemes.light?  defaultTheme :
darkTheme,
     ...
   );
 }
}
` ` `
```

PERSISTING THEME

The `switchTheme()` method toggles the `currentTheme`, and saves the currently selected theme's id in the Shared preference using key/value pair. The `switchTheme()` method is called from the `setState()` method that rebuilds the MaterialApp widget, and hence reassigns `currentTheme` to the `theme` property of the MaterialApp.

```
` ` `
class _BooksAppState extends State<BooksApp> {
 AppThemes currentTheme = AppThemes.light;
```

```
// Save theme_id using SharedPreference
Future<void> switchTheme() async {
    currentTheme =
        currentTheme == AppThemes.light?  AppThemes.dark :
AppThemes.light;

        // save current selection
        var sharedPrefs = await SharedPreferences.getInstance();
        await sharedPrefs.setInt('theme_id', currentTheme.index);
}

    @override
    Widget build(BuildContext context) {
        return MaterialApp(
            theme: currentTheme == AppThemes.light?  defaultTheme :
darkTheme,
            home: Scaffold(
                appBar: AppBar(
                    ...,
                    actions: [
                        IconButton(
                            icon: Icon(Icons.all_inclusive),
                            onPressed: () {
                                setState(() {
                                    switchTheme();
                                });
                            },
                        )
                    ]),
                body: BooksListing(),
            ),
        );
    }
}
```

SOURCE CODE ONLINE

The full source code of this example (Persisting Data: Shared preferences) is available at the GitHub.

LOCAL DATABASE (MOOR LIBRARY)

The Local database implementation uses moor library, which is based on SQLite database. In this section, you'll learn to save the preferred theme in the app's Local database to persist the last selected theme across app restarts using Moor library (Moor: Persistence library for Dart).

The Moor is the reactive persistence library for Dart and Flutter applications. It lets tables to be defined in Dart or SQL while supporting it with seamless and easy to use query API and streaming results. It persists selected themes in Flutter application's Local database using Moor plugin (moor plugin).

PACKAGE DEPENDENCIES

The following dependencies need to be added to *pubspec.yaml* configuration file.

- moor plugin (moor plugin): Persistence library built on top of *sqlite* for Dart and Flutter. It works on Android, iOS, Web platforms, and native Dart applications for persisting data in Local databases.
- moor_ffi plugin (moor_ffi plugin): This Flutter plugin generates Dart bindings to sqlite by using dart:ffi (dart:ffi library). The *'ffi'* stands for Foreign Function Interface. This plugin can be used with Flutter and/or Dart VM applications and supports all platforms where sqlite3 is installed: iOS, Android (Flutter), macOS, Linux and Windows. However, at the time of this writing Moor plugin is in mid of migration to cleaner implementation, and this plugin is being phased out. It is recommended to migrate to a newer implementation (Phasing out the moor_ffi package). The newer implementation requires sqlite3_flutter_libs (sqlite3_flutter_libs) to be added as a dependency. We'll be adding `sqlite3 _ flutter _ libs` (sqlite3_flutter_libs) dependency instead of `moor _ ffi plugin` (moor_ffi plugin).
- path_provider plugin (path_provider): This Flutter plugin is used for accessing file systems on Android and iOS platforms.
- path plugin (path): A cross-platform filesystem path manipulation library for Dart.
- moor_generator plugin (moor_generator): This library contains the generator that turns your Table classes from moor into database code.
- build_runner plugin (build_runner): This package is used to generate files. We need this package to be able to run this command `flutter packages pub run build _ runner build -delete-conflicting-outputs` to generate '*.g.dart' files.

```
` ` `
dependencies:
  moor: ^3.3.1
  sqlite3_flutter_libs: ^0.2.0
  #Helper to find the database path on mobile
  path_provider: ^1.6.11
  path: ^1.7.0

dev_dependencies:
  flutter_test:
    sdk: flutter
  build_runner: ^1.10.2
  moor_generator: ^3.3.1
` ` `
```

PREPARING DATABASE USING MOOR

First, we'll use Moor to prepare a database to save `themeId` and `theme-Name`. The active theme's id will be saved in the database table. This table will have only one entry at a given time. When the theme switches from light to dark, the older entry will be deleted, and a newly selected theme's id will be added to this table. I kept it simple on purpose to demonstrate how Moor can be integrated in your app.

The `ThemePrefs` class extends `Table`. ThemePrefs table contains only two fields: theme _ id to save id for the theme and another field themeName for saving name.

```
class ThemePrefs extends Table {
 IntColumn get themeId => integer()();
 TextColumn get themeName => text()();
}
```

It will generate a table called theme _ prefs for us. The rows of that table will be represented by a class called `ThemePref` auto generated by `moor _ generator` plugin.

Following part actually prepares the database table. This is the class where migration strategy is described. I kept the migration strategy simple on purpose. It resets the tables, and makes the light theme default in case of first launch or upgrade.

```
@UseMoor(tables: [ThemePrefs])
class MyDatabase extends _$MyDatabase {
   @override
   MigrationStrategy get migration {...}
}
```

There is one method to activate the theme. It adds the current theme's index/id to the table.

```
void activateTheme(AppThemes theme) {
 ThemePref pref =
     ThemePref(themeId: theme.index, themeName: theme.
toString());
 into(themePrefs).insert(pref);
}
```

The other method `deactivateTheme(int i)` removes the entry from the table for the given `theme _ id`.

```
` ` `
void deactivateTheme(int i) =>
   (delete(themePrefs)..where((t) => t.themeId.equals(i))).
go();
` ` `
```

The method `themeIdExists(.)` checks if the entry for given `theme _ id` already exists, and returns a boolean.

```
` ` `
Stream<bool> themeIdExists(int id) {
 return select(themePrefs)
     .watch()
     .map((prefs) => prefs.any((pref) => pref.themeId == id));
}
` ` `
```

The `getActiveTheme()` queries the table and returns the only available row. Remember there's only one row for the active theme in the whole table. By the way, this may not be the good use of a database to just store one entry. I chose to keep this way to understand the database integration in a Flutter app.

```
` ` `
Future<ThemePref> getActiveTheme() {
   return select(themePrefs).getSingle();
}
` ` `
```

Let's take a look at the database file: *themes_pref.dart* below:

```
` ` `
import 'package:moor/moor.dart';
import '../themes.dart';
part 'theme_prefs.g.dart';

// It will generate a table called "theme_prefs" for us. The
//rows of that table will be represented by a class called
//"ThemePref".
class ThemePrefs extends Table {
 // AppThemes id
 IntColumn get themeId => integer()();
 TextColumn get themeName => text()();
}

// Moor prepares database table
@UseMoor(tables: [ThemePrefs])
class MyDatabase extends _$MyDatabase {
 MyDatabase(QueryExecutor e) : super(e);

 // Bump schemaVersion whenever there's change.
 @override
```

```
int get schemaVersion => 1;

//Keeping it simple
//reset the database whenever there's an update.
// Add light theme as default theme after first launch and
upgrade
@override
MigrationStrategy get migration {
  return MigrationStrategy(onCreate: (Migrator m) {
    return m.createAll();
  }, onUpgrade: (Migrator m, int from, int to) async {
    m.deleteTable(themePrefs.actualTableName);
    m.createAll();
  }, beforeOpen: (details) async {
    if (details.wasCreated) {
      await into(themePrefs).insert(ThemePrefsCompanion(
        themeId: const Value(0),
        themeName: Value(AppThemes.light.toString()),
      ));
    }
  });
}

void activateTheme(AppThemes theme) {
  ThemePref pref =
      ThemePref(themeId: theme.index, themeName:
theme.toString());
  into(themePrefs).insert(pref);
}

void deactivateTheme(int i) =>
    (delete(themePrefs)..where((t) =>
t.themeId.equals(i))).go();

//The stream will automatically emit new items whenever the
underlying data changes.
Stream<bool> themeIdExists(int id) {
  return select(themePrefs)
      .watch()
      .map((prefs) => prefs.any((pref) => pref.themeId == id));
}

Future<ThemePref> getActiveTheme() {
  return select(themePrefs).getSingle();
}
}
}
```

Please note that this line 'part 'theme _ prefs.g.dart';' will show an error in the beginning because this file doesn't exist yet. You'll need to execute following command to generate sqlite bindings:

```
` ` `
flutter packages pub run build_runner build
--delete-conflicting-outputs
` ` `
```

CROSS-PLATFORM DATABASE IMPLEMENTATION(S)

Different platforms have different implementations for databases. We need to create a shared code that can pull in the correct database implementation for a given platform. We'll create one file to write shared code, say *shared.dart*. This file is responsible for picking the platform-specific database implementation.

File: *shared.dart*

```
` ` `
export 'unsupported.dart'
   if (dart.library.html) 'web.dart'
   if (dart.library.io) 'native.dart';
` ` `
```

When the Flutter application is running on Android, iOS, desktop (Linux/Windows/ MacOS) platforms, *mobile.dart* implementation is picked. For Flutter Web, *web.dart* is chosen by the platform. The *unsupported.dart* implementation is picked for everything else.

File: *native.dart*

```
` ` `
import 'dart:io';
import 'package:moor/ffi.dart';
import 'package:moor/moor.dart';
import 'package:path/path.dart' as p;
import 'package:path_provider/path_provider.dart' as paths;
import '../theme_prefs.dart';

//Note: Implementation borrowed from template project: //
https://github.com/appleeducate/moor_shared
MyDatabase constructDb({bool logStatements = false}) {
  if (Platform.isIOS || Platform.isAndroid) {
    final executor = LazyDatabase(() async {
      final dataDir = await paths.
getApplicationDocumentsDirectory();
      final dbFile = File(p.join(dataDir.path, 'db.sqlite'));
      return VmDatabase(dbFile, logStatements: logStatements);
    });
    return MyDatabase(executor);
  }
```

```
 if (Platform.isMacOS || Platform.isLinux) {
   final file = File('db.sqlite');
   return MyDatabase(VmDatabase(file, logStatements:
logStatements));
 }
 if (Platform.isWindows) {
   final file = File('db.sqlite');
   return MyDatabase(VmDatabase(file, logStatements:
logStatements));
 }
 return MyDatabase(VmDatabase.memory(logStatements:
logStatements));
}
~ ~ ~
```

File: *web.dart*

```
~ ~ ~

import 'package:moor/moor_web.dart';
import '../theme_prefs.dart';

MyDatabase constructDb({bool logStatements = false}) {
 return MyDatabase(WebDatabase('db', logStatements:
logStatements));
}
~ ~ ~
```

File: *unsupported.dart*

```
~ ~ ~

import '../theme_prefs.dart';
MyDatabase constructDb({bool logStatements = false}) {
 throw 'Platform not supported';
}
~ ~ ~
```

LOADING THEME

The `loadActiveTheme()` method is called from `initState()` during the app start-up time. It calls the `loadActiveTheme()` method to retrieve persisted `theme _ id` from the database. If no entry is found in the database, then the default theme is applied. The `currentTheme` is updated in `setState` method to rebuild the MaterialApp to apply the latest currentTheme.

```
~ ~ ~

// Fetching theme_id DB
void loadActiveTheme(BuildContext context) async {
 ThemePref themePref = await _database.getActiveTheme();
 setState(() {
```

```
    currentTheme = AppThemes.values[themePref.themeId];
  });
}
```
` ` `

PERSISTING THEME

The `switchTheme()` method toggles previously selected theme `oldTheme`.
The `oldTheme` is removed from the database using `deactivateTheme(.)`.
Newly updated currentTheme is added to the database using `activateTh-
eme(...)`. The `activateTheme(...)` method is called from the `setState`
method to force rebuild the MaterialApp and update the `theme` property to
the latest currentTheme.

` ` `

```
// Save theme_id in DB
Future<void> switchTheme() async {
 var oldTheme = currentTheme;

 currentTheme == AppThemes.light
     ? currentTheme = AppThemes.dark
     : currentTheme = AppThemes.light;

 //check if theme_id entry exists in table already
 var isOldThemeActive = _database.themeIdExists(oldTheme.
index);
 //Only active theme id is present in the db.
 // Remove any existing theme id from DB before adding new
entry
 if (isOldThemeActive != null) {
   _database.deactivateTheme(oldTheme.index);
 }
 setState(() {
   _database.activateTheme(currentTheme);
 });
}
```
` ` `

SOURCE CODE ONLINE

The full source code of this example (Persisting Data: Moor Library) is available at
the GitHub.

LIGHT THEME ON MULTIPLE PLATFORMS

This section demonstrates how light theme looks for *BooksApp* across multiple
platforms.

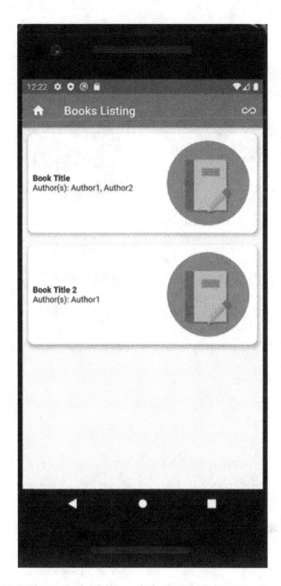

FIGURE 11.3 Light theme on Android

ANDROID

Light theme preference retrieved from disk on Android platform (Figure 11.3).

iOS

Light theme preference retrieved from disk on iOS platform (Figure 11.4).

FIGURE 11.4   Light theme on iOS

WEB

Light theme preference retrieved from disk on Web platform (Figure 11.5).

DESKTOP (macOS)

Persisted light theme preference retrieved from disk on Desktop (macOS) platform (Figure 11.6).

DARK THEME ON MULTIPLE PLATFORMS

This section demonstrates how dark theme renders for *BooksApp* across multiple platforms.

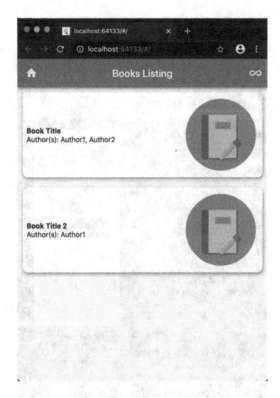

FIGURE 11.5 Light theme on Web

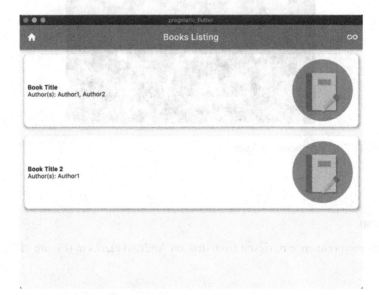

FIGURE 11.6 Light theme on Desktop (macOS)

FIGURE 11.7   Dark theme on Android

Dark theme preference retrieved from disk on Android platform (Figure 11.7).

iOS

Persisted Dark theme applied to BooksApp on iOS platform (Figure 11.8).

FIGURE 11.8 Dark theme on iOS

FIGURE 11.9   Dark theme on Web

Web

Persisted Dark theme applied to BooksApp on the Web platform (Figure 11.9).

Desktop (macOS)

Persisted dark theme preference retrieved from disk on Desktop (macOS) platform (Figure 11.10).

CONCLUSION

In this chapter, you learned to use Shared preferences and Local databases to store and retrieve data from the Flutter applications across multiple platforms. Shared preferences are implemented using Shared preference Flutter plugin. The

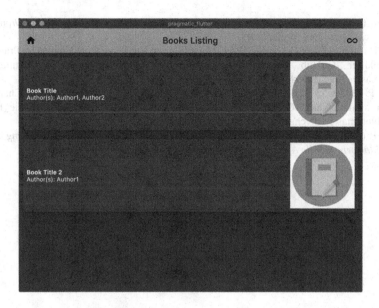

FIGURE 11.10   Dark theme on Desktop (macOS)

data saved in Local database using the moor library, a wrapper around the sqlite database.

REFERENCES

Apple. (2020, 11 22). *NSUserDefaults*. Retrieved from developer.apple.com: https://developer.apple.com/documentation/foundation/nsuserdefaults

Binder, S. (2020, 07 23). *sqlite3_flutter_libs*. Retrieved from Pub.dev: https://pub.dev/packages/sqlite3_flutter_libs

Binder, S. (2020, 11 20). *Moor: Persistence library for Dart*. Retrieved from https://moor.simonbinder.eu/

Binder, S. (2020, 11 22). *moor plugin*. Retrieved from simonbinder.eu: https://pub.dev/packages/moor

Binder, S. (2020, 11 22). *moor_ffi plugin*. Retrieved from pub.dev: https://pub.dev/packages/moor_ffi

Binder, S. (2020, 11 22). *moor_generator*. Retrieved from pub.dev: https://pub.dev/packages/moor_generator

Binder, S. (2020, 11 22). *Phasing out the moor_ffi package*. Retrieved from GitHub Issues: https://github.com/simolus3/moor/issues/691

dart.dev. (2020, 11 22). *build_runner*. Retrieved from pub.dev: https://pub.dev/packages/build_runner

dart.dev. (2020, 11 22). *path*. Retrieved from pub.dev: https://pub.dev/packages/path

flutter.dev. (2020, 11 22). *path_provider*. Retrieved from Pub.Dev: https://pub.dev/packages/path_provider

Google. (2020, 11 22). *dart:ffi library*. Retrieved from Dart Documentation: https://api.dart.dev/stable/2.7.0/dart-ffi/dart-ffi-library.html

Google. (2020, 11 22). *SharedPreferences*. Retrieved from developer.android.com: https://developer.android.com/reference/android/content/SharedPreferences

pub.dev. (2020, 11 03). *Shared preferences plugin*. Retrieved from pub.dev: https://pub.dev/packages/shared_preferences

Tyagi, P. (2020, 11 22). *Persisting Data: Moor Library*. Retrieved from Chapter11: Pragmatic Flutter GitHub Repo: https://github.com/ptyagicodecamp/pragmatic_flutter/tree/master/lib/chapter11/db

Tyagi, P. (2020, 11 22). *Persisting Data: Shared preferences*. Retrieved from Chapter11: Pragmatic Flutter GitHub Repo: https://github.com/ptyagicodecamp/pragmatic_flutter/blob/master/lib/chapter11/main_11_sharedprefs.dart

Tyagi, P. (2021). Chapter 10: Flutter Themes. In P. Tyagi, *Pragmatic Flutter: Building Cross-Platform Mobile Apps for Android, iOS, Web & Desktop*. CRC Press.

12 Integrating REST API

This chapter is an introduction to fetching data from a remote REST (Representational state transfer) API (Application Programming Interface) in a Flutter app. The Representational state transfer is a software architectural style for an API that uses less bandwidth for data transfer. It uses HTTP requests to access data. You'll learn how to access The Google Books API (Books API v1 (Experimental)) to fetch books listing for the given criteria. The Google Books API allows to view books' data in JSON representation (JSON Representation) over the HTTP. The JavaScript Object Notation (JSON) is a type of data interchange format. The JSON format is a programming language that is independent and text-based. It uses key/value pairs to store information and is human-readable.

Google Books (Google Books) is an effort to digitize the world's books. The Google Books API lets developers search books based on content. We will use this API to search books that match specific criteria and fetch book listings using this REST API. Once we have data available, we will render the raw JSON formatted data in a simple Flutter user interface. We will touch base on setting up an API key on Google API Console, and use Books API. We will dive into API details and learn how to make a REST API call to fetch results and display it in a Flutter app.

By the end of this chapter, you'll have a good understanding of getting your own API key from Google API Console. You will use this API key to make a REST call to fetch book listings. Finally, you will learn to display the raw JSON response returned from API in a Flutter widget.

WHAT IS AN API?

An API (API) is an acronym for Application Programming Interface. An API lets two applications talk to each other. We will use API to fetch book information to learn to consume APIs in Flutter applications.

The Google Books API v1 (Books API v1 (Experimental)) provides programmatic access to content and operations available on Google Books Website (Google Books). Developers can use this API to build creative reading applications using Google Books data. Google Books API is v1 and experimental at the time of this writing and provides the following features:

- Searching and browsing a list of books for specified criteria.
- Viewing book information like metadata, sales information, and preview links.
- Access to users' bookshelves and help manage their own bookshelves.

The Books API contains information about digitized books available through the Google Books Website. API is the way to facilitate the app to fetch information from this giant book database into our application. At the time of this writing, the Google

Books API has version v1 and is still experimental. This API provides a generous-free daily API quota limit of 1,000 queries per day. This is good enough to learn to use this API and build an application around it. In our *BooksApp,* we'll be fetching public book-related data using Google Books API (Books API v1 (Experimental)).

FLUTTER CONFIGURATION

We need to enable Flutter to be able to make HTTP requests. We'll use the http (http) package, a composable, Future-based library for making HTTP requests.

ADDING PACKAGE IN *pubspec.yaml*

Add the `http` package under the dependencies section. Be careful about indenting the YAML file. At the time of writing, the package version is 0.12.1. Please update the package to the latest version, if available.

```
dependencies:
  flutter:
    sdk: flutter
  http: ^0.12.1
```

IMPORTING THE PACKAGE

We'll be making HTTP requests from code to fetch remote job data. To do so, we need Dart's http package available to the code. Import the http package in your code. Use the library prefix (Effective Dart: Style) along with `as` for readability. Using a library prefix `http` is useful to write readable and intuitive code. You can call methods from the `http.dart` package on the `http` library prefix instead.

```
import 'package:http/http.dart' as http;
```

API KEY

An API key is a unique identifier given to a user, application, or developer to be able to make API requests. Requests to fetch public data from Google Books API need to be accompanied by either an API key or an access token. It is used to track the number of requests made by the app and to verify if the requests are being made from the authorized app.

In the code, we'll be using a variable `apiKey` to store the API key to access the Google Books API.

```
final apiKey = "YOUR_API_KEY";
```

First, you'll need to get an API key for yourself. You need this to be able to make API requests to Google servers. Once you have your own API key, don't forget to replace it with `"YOUR _ API _ KEY"`.

GETTING AN API KEY

Our *BooksApp* makes requests to fetch public data from Books API. Such requests need to be accompanied by an identifier like an API key (Using API keys). Follow these steps to request an API key for Google Books API:

- The first step is to acquire the API key (Acquiring and using an API key) on Google Console's credentials page (Credentials). You need to have a Google account to be able to log in and create a project.
- Create a project in Google Cloud if you don't have an existing project yet.
- Click on 'Create Credentials' -> 'API Key'
- Clicking on 'API Key' will generate the key and will provide an option to restrict key access to the chosen APIs. It's a good idea to restrict API access to Books API only (Figure 12.1).
- Your API key will be available under the 'API keys' section on the 'Credentials' page. You may want to rename the key to something memorable for future references.
- Keep this API key safe to be used in the code.
- Update the API key in the code.

ACCESSING API

Now that you have your API key keep it safe & handy to use in your code later. The placeholder "YOUR_API_KEY" shall be updated with your API key. Google Books API gives a generous-free API quota of a limit of 1,000 queries per day to explore and learn API.

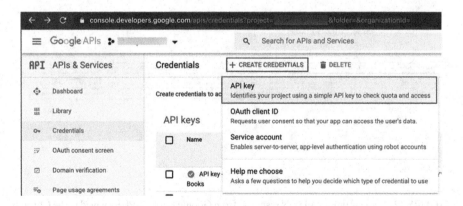

FIGURE 12.1   Creating credentials in the Google Cloud console interface

API ENDPOINT

An endpoint is a place where the resource lives and where an API sends requests to. Usually, an endpoint is a URL of the service. The API endpoint or URL to fetch the listing of Python's programming books is as below. You would need to replace `apiKey` with the API keys generated in Google Cloud Console earlier.

```
https://www.googleapis.com/books/v1/volumes?key=apiKey&q=pytho
n±programming
```

Your endpoint is ready to make an HTTP request. Try to copy and paste this URL in the browser of your choice, and observe the response displayed in the browser window. You'll observe a big blob of text in JSON format displayed in the browser's display area. Refer to Figure 12.2 to see how the response from API is rendered in the Chrome browser. Chrome is used as the browser of choice throughout this book.

Make HTTP Request

At this point, you're ready to make a REST API call to fetch book listing from your Flutter app. Let's check out the code to make an HTTP call to Google Books API. The package `http` imported earlier is used to make the HTTP request in the code snippet below. Let's create a function `makeHttpCall()` to make such calls. The `http.get(...)` method returns a `Future` object. That's why it must be called from an `async` function. This is the reason to mark `makeHttpCall()` function with `async` keyword.

The `await` keyword is used to make network calls asynchronously without blocking the main thread. The following code returns the API response of the HttpResponse (HttpResponse class) type. It will print this text on the console: `flutter: Instance of 'Response'`.

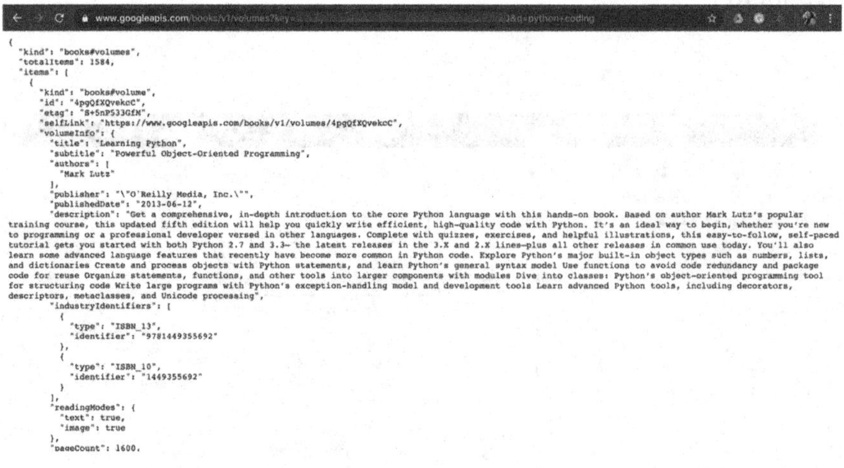

FIGURE 12.2   Raw dump of JSON response from Google Books API displayed in the Chrome browser

```
```
//Making HTTP request
Future<String> makeHttpCall() async {
  final apiKey = "$YOUR_API_KEY";
  final apiEndpoint =
"https://www.googleapis.com/books/v1/volumes?key=$apiKey&q=pyt
hon+programming";
  final http.Response response =
      await http.get(apiEndpoint, headers: {'Accept':
'application/json'});

  //This will print 'flutter: Instance of 'Response'' on
console.
  print(response);
  return response.body;
}
```
```

The `response.body` will return the body of the response, which is the actual book listing information that we're interested in.

BUILDING SIMPLE INTERFACE

Let's create a simple, scrollable interface that can accommodate the large text blob. Let's create a simple Flutter app say, *BooksApp,* which makes a REST API call to fetch book listing data from Books API and later displays this big blob of text in a scrollable interface.

Let's name the app's root widget as BooksApp, which extends the StatelessWidget widget. The BooksApp widget is like a container for the app, which provides the MaterialApp scaffolding to the app. The `debugShow-CheckedModeBanner` is used to disable the `debug` banner at the top-right corner of the app. You can leave it on if that's your preference. The `home` attribute assigns the homepage for the app.

BooksApp WIDGET

```
```
class BooksApp extends StatelessWidget {
  @override
  Widget build(BuildContext context) {
    return MaterialApp(
      debugShowCheckedModeBanner: false,
      home: BookListing(),
    );
  }
}
```
```

Next, we'll work on BookListing, a StatefulWidget used as the homepage.

BooksListing WIDGET

The *BooksApp* has BookListing stateful widget as its home property. This widget actually makes the REST call to fetch jobs data. Since remote data is fetched asynchronously, it may not be available at the time of the building widget. The app may need to update the displayed data later on. The BookListing widget is a StatefulWidget. The StatefulWidget helps to rebuild the interface whenever data is changed/updated. It uses the `setState()` method to update the data. The interface rebuilds whenever data is changed inside the `setState()` method.

The widget's state is managed by the `_BookListingState` class. The `_BookListingState` defines the method to fetch book listings as the `fetchBooks()` method. This is marked `async` to support making API calls asynchronously without blocking the app. The `build()` method makes the REST call using the `fetchBooks()` method and then goes on building its interface.

The variable `booksResponse` of type `String` holds the data fetched from the HTTP call.

The `fetchBooks()` method makes the remote call via `makeHttpCall()` and receives data over the network asynchronously. It stores network response in variable `response`. Later, `booksResponse` is updated with `response` in `setState()` method, and `build()` method is called to rebuild the interface again.

```
```
class BookListing extends StatefulWidget {
  @override
  _BookListingState createState() => _BookListingState();
}

class _BookListingState extends State<BookListing> {
  String booksResponse;

  //method to fetch books asynchronously
  fetchBooks() async {
    //making REST API call
    var response = await makeHttpCall();

    //Updating booksResponse to fetched remote data
    setState(() {
      booksResponse = response;
    });
  }

  @override
  Widget build(BuildContext context) {
    //fetching books listing
    fetchBooks();
    return Scaffold(
      body: SingleChildScrollView(
        child: booksResponse != null
```
```

```
            ? Text("Google Books API response\n $booksResponse")
            : Text("No Response from API"),
      ),
    );
  }
}
```

LOADING DATA AT APP STARTUP

As you can see that `build()` method is triggered whenever `setState()` is updated.

However, it may not be a wise choice because every time the interface is rebuilt, `fetchBooks()` will be called, and an API request in turn. It can quickly go in a cycle and end up making a network request every time. This could lead to numerous amounts of API calls to Google Books API and can run out your free API quota limit quickly and/or can incur API request charges needlessly.

To solve this problem, it makes sense to make sure that you make REST API call only whenever needed. In our case, we need to fetch data only once. To do so, it makes sense to call `initState()` method from ` _ BookListingState`. This method is executed only one time in the lifecycle of the `StatefulWidget`.

```
class _BookListingState extends State<BookListing> {
  String booksResponse;

  //method to fetch books asynchronously
  fetchBooks() async {
    //making REST API call
    var response = await makeHttpCall();

    //Updating booksResponse to fetched remote data
    setState(() {
      booksResponse = response;
    });
  }

  @override
  void initState() {
    super.initState();
    fetchBooks();
  }

  @override
  Widget build(BuildContext context) {
    return Scaffold(
      body: SingleChildScrollView(
        child: booksResponse != null
            ? Text("Google Books API response\n $booksResponse")
            : Text("No Response from API"),
```

```
      ),
    );
  }
}
` ` `
```

Now that we've remote data fetched and updated in the `booksResponse` variable, we are ready to display it in our interface. Since this data is a large blob of text, we need a widget that can adapt to scrolling if needed. The `SingleChildScrollView` widget can help with this requirement. It can accommodate a large blob of text in a scrolling manner.

As soon as we start the app, it displays a message: 'No Response from API' on-screen since no data is available yet. The remote data is being fetched using the `await makeHttpCall()` method. The variable `booksResponse` is updated in the `setState()` method. The `setState` method triggers rebuilding the interface with newly fetched data, and a large text blob is displayed on the screen.

RUNNING CODE

Let's put what we have learned so far together and run the code to see the Flutter application displaying data on the screen on all four platforms: Android, iOS, MacOS, and Chrome.

Note: Don't forget to update the apikey with your own key. Replace "YOUR_API_KEY" with the key you obtained earlier.

COMPLETE CODE

```
` ` `
import 'package:flutter/material.dart';
import 'package:http/http.dart' as http;
import '../config.dart';

//Making HTTP request
Future<String> makeHttpCall() async {
  final apiKey = "$YOUR_API_KEY";
  final apiEndpoint =
      "https://www.googleapis.com/books/v1/
volumes?key=$apiKey&q=python";
  final http.Response response =
      await http.get(apiEndpoint, headers: {'Accept':
'application/json'});

  //This will print `flutter: Instance of 'Response'` on
console.
  print(response);
  return response.body;
}
```

```dart
class BooksApp extends StatelessWidget {
  @override
  Widget build(BuildContext context) {
    return MaterialApp(
      debugShowCheckedModeBanner: false,
      home: BookListing(),
    );
  }
}


class BookListing extends StatefulWidget {
  @override
  _BookListingState createState() => _BookListingState();
}


class _BookListingState extends State<BookListing> {
  String booksResponse;

  //method to fetch books asynchronously
  fetchBooks() async {
    //making REST API call
    var response = await makeHttpCall();

    //Updating booksResponse to fetched remote data
    setState(() {
      booksResponse = response;
    });
  }

  @override
  void initState() {
    super.initState();
    fetchBooks();
  }

  @override
  Widget build(BuildContext context) {
    //fetching books listing
    //fetchBooks();

    return Scaffold(
      body: SingleChildScrollView(
        child: booksResponse != null
            ? Text("Google Books API response\n $booksResponse")
            : Text("No Response from API"),
      ),
    );
  }
}
```

MacOS Target

You'll need to add an entitlement, com.apple.security.network.client (com.apple.security.network.client) in order to make any network request.

Note: Using App Sandbox (App Sandbox) is required if you plan to distribute your application in the App Store.

ADDING ENTITLEMENT

The macOS builds are signed by default and sandboxed with App Sandbox. Add the network entitlement in *macos/Runner/DebugProfile.entitlements* and *macos/Runner/Release.entitlements* as below:

```
```
<key>com.apple.security.network.client</key>
```
```

Figure 12.3 shows the API response displayed at desktop (macos).

Android Target

The following screenshot is taken on the *Pixel 3 API 28* emulator.

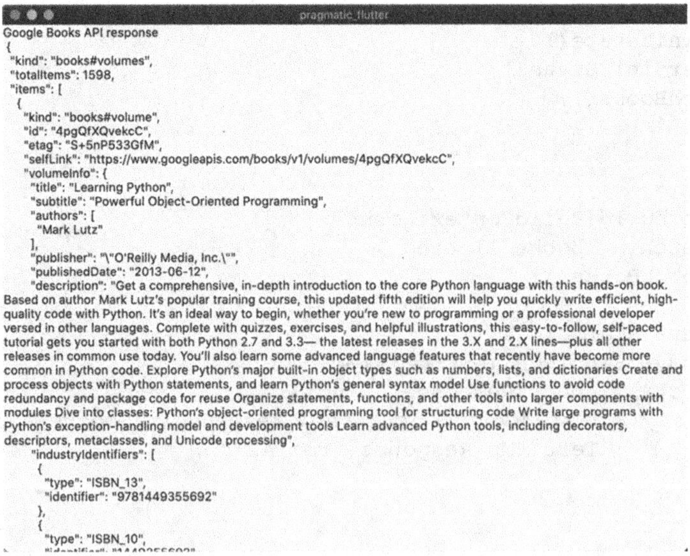

Google Books API response
```
{
  "kind": "books#volumes",
  "totalItems": 1598,
  "items": [
    {
      "kind": "books#volume",
      "id": "4pgQfXQvekcC",
      "etag": "S+5nP533GfM",
      "selfLink": "https://www.googleapis.com/books/v1/volumes/4pgQfXQvekcC",
      "volumeInfo": {
        "title": "Learning Python",
        "subtitle": "Powerful Object-Oriented Programming",
        "authors": [
          "Mark Lutz"
        ],
        "publisher": "\"O'Reilly Media, Inc.\"",
        "publishedDate": "2013-06-12",
        "description": "Get a comprehensive, in-depth introduction to the core Python language with this hands-on book. Based on author Mark Lutz's popular training course, this updated fifth edition will help you quickly write efficient, high-quality code with Python. It's an ideal way to begin, whether you're new to programming or a professional developer versed in other languages. Complete with quizzes, exercises, and helpful illustrations, this easy-to-follow, self-paced tutorial gets you started with both Python 2.7 and 3.3— the latest releases in the 3.X and 2.X lines—plus all other releases in common use today. You'll also learn some advanced language features that recently have become more common in Python code. Explore Python's major built-in object types such as numbers, lists, and dictionaries Create and process objects with Python statements, and learn Python's general syntax model Use functions to avoid code redundancy and package code for reuse Organize statements, functions, and other tools into larger components with modules Dive into classes: Python's object-oriented programming tool for structuring code Write large programs with Python's exception-handling model and development tools Learn advanced Python tools, including decorators, descriptors, metaclasses, and Unicode processing",
        "industryIdentifiers": [
          {
            "type": "ISBN_13",
            "identifier": "9781449355692"
          },
          {
            "type": "ISBN_10",
```

FIGURE 12.3   API response displayed in macos target

ANDROIDMANIFEST.XML

Add the following permission to enable Internet access. This permission is not required in debug mode.

```
<uses-permission android:name="android.permission.
INTERNET"/>
```

Figure 12.4 shows the API response displayed at Android platform.

FIGURE 12.4 API response displayed in the Android target

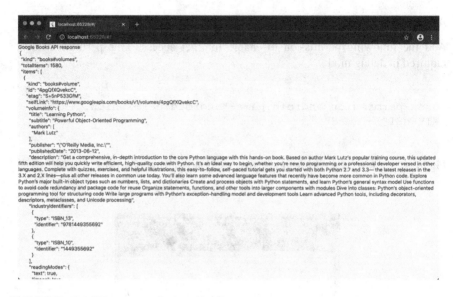

FIGURE 12.5   API response displayed in Chrome target

CHROME TARGET

Figure 12.5 shows the API response displayed at Web (Chrome) platform.

iOS TARGET

The following screenshot is taken from iPhone SE (2nd generation) simulator. This is the default simulator selected for my Xcode configuration. API response is displayed as shown in Figure 12.6.

SOURCE CODE ONLINE

Source code for this example (Calling REST API) is available at GitHub.

CONCLUSION

In this chapter, you learned to get an API key to make REST API calls to fetch remote data in JSON representation. The raw data was displayed in a simple Flutter app. You learned about making network calls, fetching data using API, and loading data exactly once during app startup.

In the next chapter (Chapter 13: Data Modeling), you will learn about parsing JSON data and create object models to better manage code.

FIGURE 12.6   API response displayed in the iOS target

REFERENCES

Apple. (2020, 11 21). *App Sandbox*. Retrieved from developer.apple.com: https://developer. apple.com/documentation/security/app_sandbox

Apple. (2020, 11 21). *com.apple.security.network.client*. Retrieved from developer.apple.com: https://developer.apple.com/documentation/bundleresources/entitlements/com_apple_ security_network_client

Dart. (2020, 11 21). *Effective Dart: Style*. Retrieved from dart.dev: https://dart.dev/guides/language/ effective-dart/style#do-name-import-prefixes-using-lowercase_with_underscores

Dart Team. (2020, 12 22). *HttpResponse class*. Retrieved from Dart Dev: https://api.dart.dev/ stable/2.7.1/dart-io/HttpResponse-class.html

Flutter Team. (2020, 11 21). *http*. Retrieved from pub.dev: https://pub.dev/packages/http

Google. (2020, 11 21). *Acquiring and using an API key*. Retrieved from Google Books API: https://developers.google.com/books/docs/v1/using#APIKey

Google. (2020, 11 21). *Books API v1 (Experimental)*. Retrieved from Google Books APIs: https://developers.google.com/books/docs/overview#books_api_v1

Google. (2020, 11 21). *Credentials*. Retrieved from Google APIs: https://console.developers. google.com/apis/credentials

Google. (2020, 11 21). *Google Books*. Retrieved from books.google.com: https://books. google.com/

Google. (2020, 11 21). *Using API keys*. Retrieved from Google Cloud: https://cloud.google. com/docs/authentication/api-keys?visit_id=637269820995156061-804882371&rd=1

JSON Representation. (2020, 11 21). Retrieved from json.org: https://www.json.org/json-en. html

Tyagi, P. (2020, 11 21). *Calling REST API*. Retrieved from Chapter12: Pragmatic Flutter GitHub Repo: https://github.com/ptyagicodecamp/pragmatic_flutter/blob/master/lib/ chapter12/main_12.dart

Tyagi, P. (2021). Chapter 13: Data Modeling. In P. Tyagi, *Pragmatic Flutter: Building Cross-Platform Mobile Apps for Android, iOS, Web & Desktop*. CRC Press.

Wikipedia. (2020, 11 21). *API*. Retrieved from Wikipedia: https://en.wikipedia.org/wiki/API

13 Data Modeling

In the previous chapter (Chapter 12: Integrating REST API), you learned to access books data from Google Books API (or Books API). You queried the API for fetching 'Python' programming books and displayed the raw blob of JSON response in a Flutter widget of *BooksApp*. Now that you know how to receive the HttpResponse (HttpResponse class) response from API. In this chapter, you will learn to convert the response into JSON and parse it to render views in the Flutter ListView (ListView class) widget. Later in this chapter, the JSON formatted response is converted into the `BookModel` data model by mapping JSON response into the data model class. These Dart objects are used to build the same book listing Flutter user interface.

PARSING JSON

The JavaScript Object Notation or JSON is a type of data interchange format. The JSON format is programming language independent and text-based. It uses key/value pairs to store information and is human-readable. Dart provides the `dart:convert` (dart:convert library) library to parse JSON data.

THE `dart:convert` LIBRARY

The `dart:convert` library parses the JSON response into the Dart collection Map. In the following code, the `apiResponse` is the response returned from `http.get(apiEndPoint)`.

The HTTP headers are set to accept JSON encoding using `` `application/json` ``. The `json.decode(...)` method takes the response's body to parse it into Dart data structures. It returns a `Map` (Map<K, V> class) as a `Future` object. Futures are the objects that return the results of the asynchronous operations.

```
```
//importing the Dart package
import 'dart:convert';

//Making HTTP request
//Function to make REST API call
Future<dynamic> makeHttpCall() async {
  //API Key: To be replaced with your key
  final apiKey = "$YOUR_API_KEY";
  final apiEndpoint = "https://www.googleapis.com/books/v1/
volumes?key=$apiKey&q=python+coding";
  final http.Response response =
  await http.get(apiEndpoint, headers: {'Accept':
'application/json'});
```
```

```
//Parsing API's HttpResponse to JSON format
//Converting string response body to JSON representation
final jsonObject = json.decode(response.body);

//Prints JSON formatted response on console
print(jsonObject);
return jsonObject;
}
```
~ ~ ~

The print(jsonObject) function prints the JSON object on the console.

The JSON response from API for one book item looks as shown in Figure 13.1. We need only some parts of the book information to show in the *BooksApp*. The parts needed for the app are highlighted in Figure 13.1. We are interested in parsing this required information only to keep things simple.

JSON FORMATTED RESPONSE

The jsonObject is a JSON encoded response from API. The JSON returned from the API looks like below. The JSON object consists of items key to hold an array of book information. The API returns about ten items at one time by default. I have omitted a few attributes to simplify the structure. Each book or item contains the following attributes/keys returned from API:

- volumeInfo
  - title: Book's title.
  - subtitle: Book's subtitle.
  - authors: Book's authors.
  - publisher: Book's publisher.
  - publishedDate: Publication date.
  - description: Book description.
  - imageLinks
    - smallThumbnail: Link to smaller sized thumbnail.
    - thumbnail: Link to thumbnail for book image.
  - saleInfo
    - saleability: Information whether a book is available for sale or not.
    - buyLink: Link to smaller sized thumbnail.
  - accessInfo
    - webReaderLink: Link to read the text in the browser.

JSON data structure is shown below for one book entry.

~ ~ ~

```
{
  "items": [
    {
      "volumeInfo": {
```

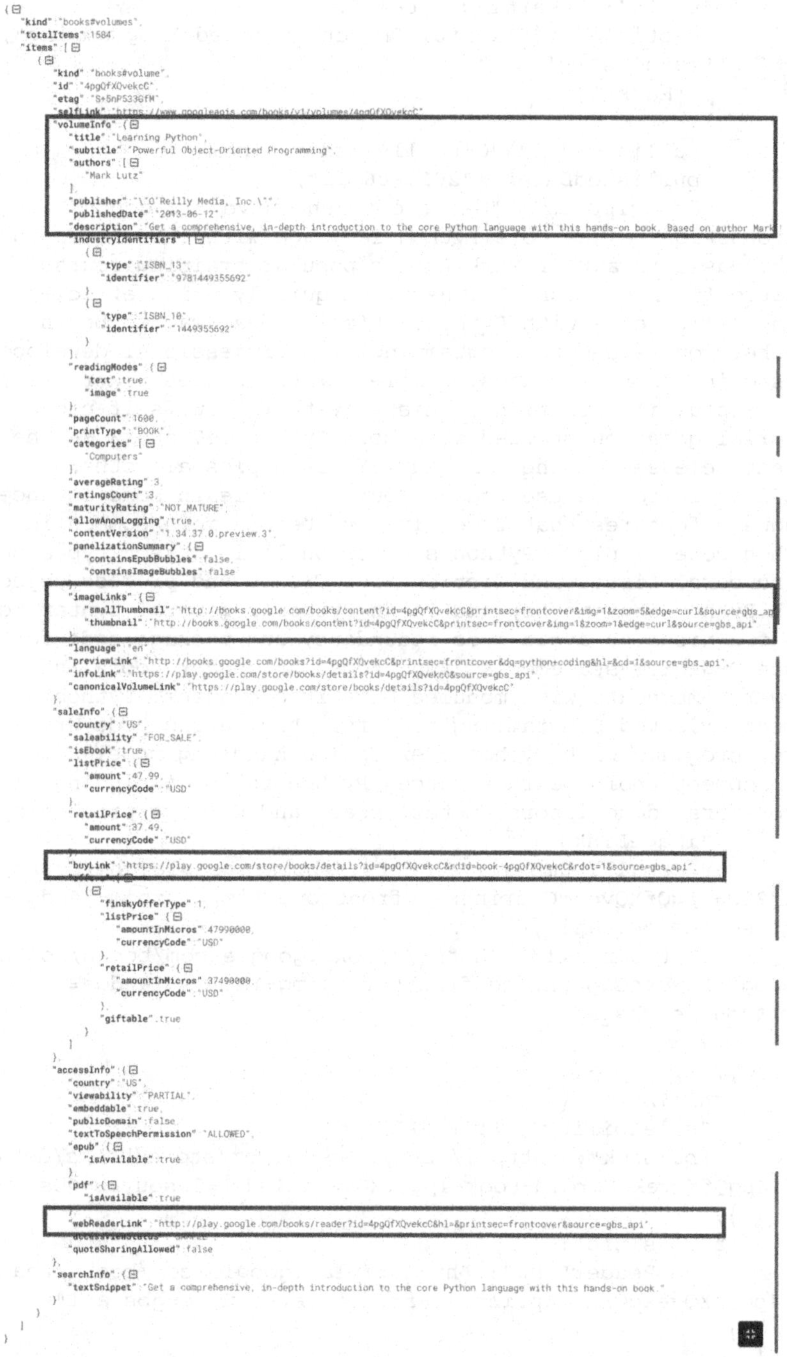

```
{
  "kind": "books#volumes",
  "totalItems": 1584,
  "items": [
    {
      "kind": "books#volume",
      "id": "4pgQfXQvekcC",
      "etag": "S+5nPS33GfM",
      "selfLink": "https://www.googleapis.com/books/v1/volumes/4pgQfXQvekcC",
      "volumeInfo": {
        "title": "Learning Python",
        "subtitle": "Powerful Object-Oriented Programming",
        "authors": [
          "Mark Lutz"
        ],
        "publisher": "\"O'Reilly Media, Inc.\"",
        "publishedDate": "2013-06-12",
        "description": "Get a comprehensive, in-depth introduction to the core Python language with this hands-on book. Based on author Mark
        "industryIdentifiers": [
          {
            "type": "ISBN_13",
            "identifier": "9781449355692"
          },
          {
            "type": "ISBN_10",
            "identifier": "1449355692"
          }
        ],
        "readingModes": {
          "text": true,
          "image": true
        },
        "pageCount": 1600,
        "printType": "BOOK",
        "categories": [
          "Computers"
        ],
        "averageRating": 3,
        "ratingsCount": 3,
        "maturityRating": "NOT_MATURE",
        "allowAnonLogging": true,
        "contentVersion": "1.34.37.0.preview.3",
        "panelizationSummary": {
          "containsEpubBubbles": false,
          "containsImageBubbles": false
        },
        "imageLinks": {
          "smallThumbnail": "http://books.google.com/books/content?id=4pgQfXQvekcC&printsec=frontcover&img=1&zoom=5&edge=curl&source=gbs_ap
          "thumbnail": "http://books.google.com/books/content?id=4pgQfXQvekcC&printsec=frontcover&img=1&zoom=1&edge=curl&source=gbs_api"
        },
        "language": "en",
        "previewLink": "http://books.google.com/books?id=4pgQfXQvekcC&printsec=frontcover&dq=python+coding&hl=&cd=1&source=gbs_api",
        "infoLink": "https://play.google.com/store/books/details?id=4pgQfXQvekcC&source=gbs_api",
        "canonicalVolumeLink": "https://play.google.com/store/books/details?id=4pgQfXQvekcC"
      },
      "saleInfo": {
        "country": "US",
        "saleability": "FOR_SALE",
        "isEbook": true,
        "listPrice": {
          "amount": 47.99,
          "currencyCode": "USD"
        },
        "retailPrice": {
          "amount": 37.49,
          "currencyCode": "USD"
        },
        "buyLink": "https://play.google.com/store/books/details?id=4pgQfXQvekcC&rdid=book-4pgQfXQvekcC&rdot=1&source=gbs_api",
        "offers": [
          {
            "finskyOfferType": 1,
            "listPrice": {
              "amountInMicros": 47990000,
              "currencyCode": "USD"
            },
            "retailPrice": {
              "amountInMicros": 37490000,
              "currencyCode": "USD"
            },
            "giftable": true
          }
        ]
      },
      "accessInfo": {
        "country": "US",
        "viewability": "PARTIAL",
        "embeddable": true,
        "publicDomain": false,
        "textToSpeechPermission": "ALLOWED",
        "epub": {
          "isAvailable": true
        },
        "pdf": {
          "isAvailable": true
        },
        "webReaderLink": "http://play.google.com/books/reader?id=4pgQfXQvekcC&hl=&printsec=frontcover&source=gbs_api",
        "accessViewStatus": "SAMPLE",
        "quoteSharingAllowed": false
      },
      "searchInfo": {
        "textSnippet": "Get a comprehensive, in-depth introduction to the core Python language with this hands-on book."
      }
    }
  ]
}
```

FIGURE 13.1 Image for JSON blob for one item

```
    "title": "Learning Python",
    "subtitle": "Powerful Object-Oriented Programming",
    "authors": [
      "Mark Lutz"
    ],
    "publisher": "\"O'Reilly Media, Inc.\"",
    "publishedDate": "2013-06-12",
    "description": "Get a comprehensive, in-depth
introduction to the core Python language with this hands-on
book. Based on author Mark Lutz's popular training course, this
updated fifth edition will help you quickly write efficient,
high-quality code with Python. It's an ideal way to begin,
whether you're new to programming or a professional developer
versed in other languages. Complete with quizzes, exercises,
and helpful illustrations, this easy-to-follow, self-paced
tutorial gets you started with both Python 2.7 and 3.3- the
latest releases in the 3.X and 2.X lines—plus all other
releases in common use today. You'll also learn some advanced
language features that recently have become more common in
Python code. Explore Python's major built-in object types such
as numbers, lists, and dictionaries Create and process objects
with Python statements, and learn Python's general syntax model
Use functions to avoid code redundancy and package code for
reuse Organize statements, functions, and other tools into
larger components with modules Dive into classes: Python's
object-oriented programming tool for structuring code Write
large programs with Python's exception-handling model and
development tools Learn advanced Python tools, including
decorators, descriptors, metaclasses, and Unicode processing",
      "imageLinks": {
        "smallThumbnail": "http://books.google.com/books/con
tent?id=4pgQfXQvekcC&printsec=frontcover&img=1&zoom=5&edge=
curl&source=gbs_api",
        "thumbnail": "http://books.google.com/books/content?
id=4pgQfXQvekcC&printsec=frontcover&img=1&zoom=1&edge=
curl&source=gbs_api"
      }
    },
    "saleInfo": {
      "saleability": "FOR_SALE",
      "buyLink": "https://play.google.com/store/books/details
?id=4pgQfXQvekcC&rdid=book-4pgQfXQvekcC&rdot=1&source=gbs_api"
    },
    "accessInfo": {
      "webReaderLink": "http://play.google.com/books/reader?
id=4pgQfXQvekcC&hl=&printsec=frontcover&source=gbs_api"
    }
  }
]
}
```

Let's re-examine the `fetchBooks()` method of `_ BooksListingState` class in the previous chapter (Chapter 12: Integrating REST API).

```
` ` `
var booksListing;
fetchBooks() async {
 var response = await makeHttpCall();
 setState(() {
   booksListing = response["items"];
 });
}
` ` `
```

The `items` property in the JSON response above holds the list of books as JSON objects.

The response is assigned to the results returned from the `makeHttpCall()`function. The `booksListing` is assigned to the `response["items"]` returned from the API. Since the `response` variable stores the JSON response above, `response["items"]` would give the list of books as a list of JSON objects. We will use this list of JSON objects to retrieve the book information and eventually render it in Flutter applications.

Note: The `response["items"]` returns a list of `LinkedHashMap` as `List<dynamic>`. LinkedHashMap (LinkedHashMap<K, V> class) is a `Hashtable` implementation of the `Map`. The `LinkedHashMap` preserves the insertion order of keys.

Each JSON object of this list contains the book information that needs to be displayed in each row of the app's books list. In our app, we're interested in displaying the title, authors, and thumbnail image of the book. We can use the following attributes from the above JSON object(s) to get that information.

* `volumeInfo->title`: title of the book.
* `volumeInfo->description`: description of the book
* `volumeInfo->imageLinks->thumbnail`: Link to thumbnail image of the book's cover page.

Now that we know how to fetch and access information about the book listing, let's start building the interface. We'll build a simple interface like below (Figure 13.2) to display the book's title, author(s), and cover image, if available.

`ListView` WIDGET: LISTING ENTRIES

In the previous chapter (Chapter 12: Integrating REST API), a big blob of API responses was displayed on the main screen. In this section, you'll see how to display each book entry in its own row using the `ListView` (ListView class) Flutter widget.

FIGURE 13.2   ListView widget listing books (generic)

`ListView` WIDGET

The `ListView` widget is used for laying out its children in a scrolling manner.

One way to build a `ListView` widget is to use `ListView.builder`. It has two main properties:

- `itemCount`: This property includes the number of items to be displayed in `ListView`.
- `itemBuilder`: This property takes an anonymous function with two parameters (context, index) to render each row of the list. The index keeps track of the number of the item in the list.

A `BuildContext` (BuildContext class) is like the handle to the location of a widget in the widget tree.

We saw earlier that `booksListing = response["items"];` holds the entries of the books returned as the API response. The `ListView.builder(...)` creates a scrollable, linear array of book item widgets that can be created on-demand. We'll be creating a `BookTile` widget to encapsulate the book item widget. This `BookTile` widget takes the current item from the `booksListing` and passes on book information for rendering in the list.

```
~~~
ListView.builder(
  itemCount: booksListing == null?  0 : booksListing.length,
  itemBuilder: (context, index) {
    //current book information passed on to BookTile
    return BookTile(book: booksListing[index]);
  },
),
~~~
```

CUSTOM WIDGET: `BookTile`

The `BookTile` widget is used to show each entry in the book listing. It consists of the book title, authors, and cover image information provided by API, along with a dividing gray line as the separator (Figure 13.3).

ANATOMY OF CUSTOM LIST ENTRY WIDGET

Let's first understand the structure of the custom list entry widget. The custom list entry widget is built using the following primitive Flutter widgets:

- `Card` Widget
- `Padding` Widget
- `Row` Widget
- `Flexible` Widget
- `Column` Widget
- `Text` Widget
- `Image` Widget

FIGURE 13.3 Custom list entry widget

The structure of the custom widget `BookTile` is shown in Figure 13.4. Since image, title, and overview text are aligned vertically, we can use the `Column` widget.

`BookTile` STATELESSWIDGET

The `BookTile` widget can be a `Stateless` widget since it creates a part of the interface and doesn't change its state afterward. Create a file *booktile.dart* to hold stateless `BookTile` widget class. The variable `` `final book` `` holds the current book information passed from the `ListView.builder()` constructor.

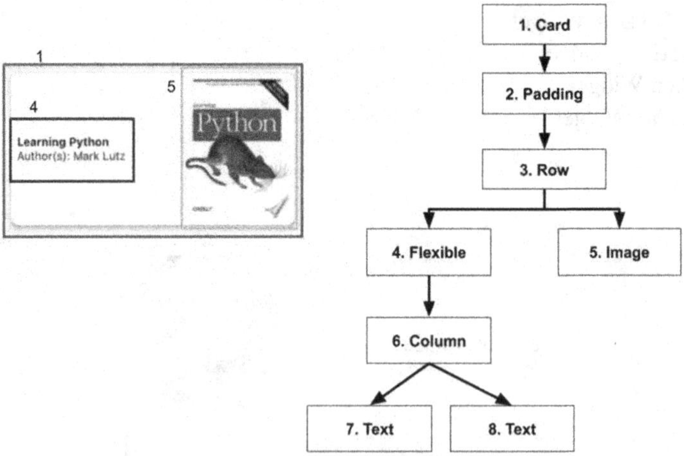

```
```
import 'package:flutter/material.dart';
class BookTile extends StatelessWidget {
 final book;
 const BookTile({Key key, this.book}) : super(key: key);
 @override
 Widget build(BuildContext context) {
     //Card widget
     //Card's child is Padding
     //Padding's child is Row
     //Row's children are Flexible and Image
     //Flexible's child is Column
     //Column's children are two Text widgets
   return Card(
     child: Padding(
       child: Row(
         children: [
           Flexible(
             child: Column(
               children: <Widget>[
                 Text(),
```

FIGURE 13.4   Anatomy of the custom list entry widget
```

```
            Text(),
          ],
        ),
      ),
      Image.network()
    ],
   ),
  ),
 );
}
}
```

Let's go over each building block widgets one by one

Card WIDGET

The Card Widget (Card class) creates a Material Design Card (Cards). It's a useful widget to show related information together. Since we want to display title, authors, and image related to one book item, it makes sense to Card widget as a list entry. It has slightly rounded corners and shadows that we can tweak. The RoundedRectangleBorder for the shape attribute is used to assign a rounded corner shape to the Card widget. The elevation can be adjusted using the elevation property. The margin property is used to give spacing around the Card widget. The Padding widget was added as its child.

```
Card(
 shape: RoundedRectangleBorder(
   borderRadius: BorderRadius.circular(10.0),
 ),
 elevation: 5,
 margin: EdgeInsets.all(10),
 child: Padding(...),
)
```

Padding WIDGET

We want little space between the Card widget edges and Row widget. We can achieve this padding around the Row widget using Padding Widget. The Padding widget is used to give padding around the Row widget.

```
Padding(
 padding: const EdgeInsets.all(8.0),
 child: Row(),
)
```

Row WIDGET

The Row widget (Row class) aligns its children in a horizontal direction. Row widget is used to display the title and author text to the left side of the screen and the book's cover page image to the right side. The mainAxisAlignment property tells Row how to place its children widgets in the available space. The MainAxisAlignment. spaceBetween helps to distribute the free space evenly between the children. The Row widget has two children to display the book's text details and image.

```
Row (
  mainAxisAlignment: MainAxisAlignment.spaceBetween,
  children: [
    Flexible(),
    Image.network()
  ],
)
```

Flexible WIDGET

The Flexible widget (Flexible class) is used to display the book's text details. It uses the Column widget as its child to align title and author information vertically. The Flexible widget gives Column widget flexibility to expand to fill the available space in the main axis. If the book's title is too long, it'll expand vertically rather than overflowing.

```
Flexible(
  child: Column()
)
```

Column WIDGET

The Column widget (Column class) aligns its children vertically. The book's title and author's information are displayed in Column widget's children. The crossAxisAlignment property is set to CrossAxisAlignment.start, which helps to place the children with their start edge aligned with the start side of the cross axis. This property makes sure that children widgets are aligned to the left of the Column widget.

```
Column (
  crossAxisAlignment: CrossAxisAlignment.start,
  children: <Widget>[
    Text(),
    Text()
  ],
)
```

Text Widgets

The Text (Text class) widgets are used to display text. Two Text widgets are placed vertically in the Column widget. The first Text widget is for the book's title text. The variable book is a parsed JSON response returned from the API for the given index. The related piece of JSON is:

```
```
"volumeInfo": {
        "title": "Learning Python",
        "authors": [
          "Mark Lutz"
          ],
}
```
```

The title is available at path 'volumeInfo'->'title'. The authors' list is available at path 'volumeInfo'->'authors'.

```
```
Text(
 '${book['volumeInfo']['title']}',
 style: TextStyle(fontSize: 14, fontWeight: FontWeight.bold),
)
```
```

The second Text widget is to display the author(s) name(s). The author's name(s) is available as a List. The authors' name list is concatenated with commas. A null check is added for authors' information to handle cases when there's no author list available.

```
```
book['volumeInfo']['authors'] != null
?   Text(
        'Author(s): ${book['volumeInfo']['authors'].join(", ")}',
        style: TextStyle(fontSize: 14),
)
    : Text(""),
```
```

Both Text widgets are styled with the same font size. However, the title is styled to be bold as well.

Image Widget

The Image widget is used to display the book's cover page image. The `Image.network()` method uses network URL to load and display images. The fit property is assigned to BoxFit.fill, which helps to fit the image in the given target box. An empty Container widget is added when there's no thumbnail information available.

```
` ` `
book['volumeInfo']['imageLinks']['thumbnail'] != null
  ? Image.network(
      book['volumeInfo']['imageLinks']['thumbnail'],
      fit: BoxFit.fill,
    )
  : Container(),
` ` `
```

FINISHED CODE (PART 1): BookTile WIDGET

The BookTile Stateless Widget is available in *booktile.dart* (Chapter 13: BookTile
(Part 1)) file available at GitHub repo.

```
` ` `
import 'package:flutter/material.dart';
class BookTile extends StatelessWidget {
  final book;
  const BookTile({Key key, this.book}) : super(key: key);
  @override
  Widget build(BuildContext context) {
    return Card(
      shape: RoundedRectangleBorder(
        borderRadius: BorderRadius.circular(10.0),
      ),
      elevation: 5,
      margin: EdgeInsets.all(10),
      child: Padding(
        padding: const EdgeInsets.all(8.0),
        child: Row(
          mainAxisAlignment: MainAxisAlignment.spaceBetween,
          children: [
            Flexible(
              child: Column(
                crossAxisAlignment: CrossAxisAlignment.start,
                children: <Widget>[
                  Text(
                    '${book['volumeInfo']['title']}',
                    style: TextStyle(fontSize: 14, fontWeight:
FontWeight.bold),
                  ),
                  book['volumeInfo']['authors'] != null
                      ? Text(
                          'Author(s):
${book['volumeInfo']['authors'].join(", ")}',
                          style: TextStyle(fontSize: 14),
                        )
                      : Text(""),
                ],
              ),
```

```
              ),
          book['volumeInfo']['imageLinks']['thumbnail'] != null
              ? Image.network(

                 book['volumeInfo']['imageLinks']
['thumbnail'],
                   fit: BoxFit.fill,
                 )
              : Container(),
          ],
        ),
      ),
    );
  }
}
```

FINISHED CODE (PART 1): MAIN METHOD

This code is available in the GitHub repo under part1 (Chapter 13: Custom Widget
(Part 1)) folder.

```
//importing the Dart package
import 'dart:convert';

import 'package:flutter/material.dart';
import 'package:http/http.dart' as http;
import 'package:pragmatic_flutter/chapter15/part1/booktile.
dart';

import '../../config.dart';

//Showing book listing in ListView
class BooksApp extends StatelessWidget {
 @override
 Widget build(BuildContext context) {
   return MaterialApp(
     debugShowCheckedModeBanner: false,
     home: BooksListing(),
   );
 }
}

//Making HTTP request
//Function to make REST API call
Future<dynamic> makeHttpCall() async {
 //API Key: To be replaced with your key
 final apiKey = "$YOUR_API_KEY";
```

```
final apiEndpoint =  "https://www.googleapis.com/books/v1
/volumes?key=$apiKey&q=python+coding";
final http.Response response =
await http.get(apiEndpoint, headers: {'Accept': 'application/
json'});

//Parsing API's HttpResponse to JSON format
//Converting string response body to JSON representation
final jsonObject = json.decode(response.body);

//Prints JSON formatted response on console
print(jsonObject);
return jsonObject;
}

class BooksListing extends StatefulWidget {
@override
_BooksListingState createState() => _BooksListingState();
}

class _BooksListingState extends State<BooksListing> {
var booksListing;
fetchBooks() async {
  var response = await makeHttpCall();
  setState(() {
    booksListing = response["items"];
  });
}
@override
void initState() {
  super.initState();
  fetchBooks();
}

@override
Widget build(BuildContext context) {
  return Scaffold(
    appBar: AppBar(
      title: Text("Books Listing"),
    ),
    body: ListView.builder(
      itemCount: booksListing == null?  0 : booksListing.
length,
      itemBuilder: (context, index) {
        return BookTile(book: booksListing[index]);
      },
    ),
  );
}
}
```

CONSTRUCTING DATA MODEL

In this section, you'll learn to create a data model object from a JSON formatted API response. We will create a `BookModel` Dart class to parse the JSON response returned from API.

    `BookModel` class will have members for each of the `response["items"]` JSON attributes returned as API response.

REVISITING BOOKS API RESPONSE STRUCTURE

The JSON response returned from Books API looks like a huge blob of string. It could be hard to access the attributes by calling their names every time from code. A spelling mistake can make debugging very hard.

    To avoid this problem, it makes sense to create a Dart object mapped to the response. We will call this class `BookModel`.

    Earlier, if you would want to access the 'title' of the book, you had to access it as `response["items"]["volumeInfo"]["title"]`.

    After mapping data to the BookModel class, you can access 'title' as `bookModelObj.volumeInfo.title` and so on.

GOOGLE BOOKS API JSON RESPONSE

```
```
{
  "items": [
    {
      "volumeInfo": {
        "title": "Learning Python",
        "subtitle": "Powerful Object-Oriented Programming",
        "authors": [
          "Mark Lutz"
        ],
        "publisher": "\"O'Reilly Media, Inc.\"",
        "publishedDate": "2013-06-12",
        "description": "Get a comprehensive, in-depth
introduction to the core Python language with this hands-on
book.",
        "imageLinks": {
          "smallThumbnail": "http://books.google.com/books/
content?id=4pgQfXQvekcC&printsec=frontcover&img=1&zoom=5&edge=
curl&source=gbs_api",
          "thumbnail": "http://books.google.com/books/content?
id=4pgQfXQvekcC&printsec=frontcover&img=1&zoom=1&edge=
curl&source=gbs_api"
        }
      },
      "saleInfo": {
        "saleability": "FOR_SALE",
```
```

```
        "buyLink": "https://play.google.com/store/books/detail
s?id=4pgQfXQvekcC&rdid=book-4pgQfXQvekcC&rdot=1&source=gbs_
api"
        },
        "accessInfo": {
        "webReaderLink": "http://play.google.com/books/reader?
id=4pgQfXQvekcC&hl=&printsec=frontcover&source=gbs_api"
        }
    }
  ]
}
```

CONSTRUCTING BookModel

Let's create a BookModel class to hold the API movie listing data set. The BookModel class will have members for each attribute of the JSON response.

THE BookModel.fromJson(...)

The BookModel.fromJson(Map<String, dynamic> json) takes a JSON map and assigns corresponding values to the BookModel object's members. This is the entry point method that is called to parse JSON responses received from the network. In this case, JSON response attributes are tiered. Each tier will have its own models. The BookModel class needs access to volumeInfo, accessInfo, and saleInfo attributes/members.

BookModel CLASS

BookModel.fromJson() uses a factory constructor. The factory constructor is used when implementing a constructor that doesn't always create a new instance of its class. It's useful when getting an object from a cache rather than creating a duplicate.

```
`  `
class BookModel {
  final VolumeInfo volumeInfo;
  final AccessInfo accessInfo;
  final SaleInfo saleInfo;

  BookModel({this.volumeInfo, this.accessInfo, this.saleInfo});

  factory BookModel.fromJson(Map<String, dynamic> json) {
    return BookModel(
        volumeInfo: VolumeInfo.fromJson(json['volumeInfo']),
        accessInfo: AccessInfo.fromJson(json['accessInfo']),
        saleInfo: SaleInfo.fromJson(json['saleInfo']));
  }
}
`  `
```

VolumeInfo Class

The VolumeInfo class has a title, authors as its members. It has a reference
to the ImageLinks class to access thumbnail information.

```
```
class VolumeInfo {
 final String title;
 final String subtitle;
 final String description;
 final List<dynamic> authors;
 final String publisher;
 final String publishedDate;
 final ImageLinks imageLinks;

 VolumeInfo(
     {this.title,
     this.subtitle,
     this.description,
     this.authors,
     this.publisher,
     this.publishedDate,
     this.imageLinks});

 factory VolumeInfo.fromJson(Map<String, dynamic> json) {
   return VolumeInfo(
       title: json['title'],
       subtitle: json['subtitle'],
       description: json['description'],
       authors: json['authors'] as List,
       publisher: json['publisher'],
       publishedDate: json['publishedDate'],
       imageLinks: ImageLinks.fromJson(json['imageLinks']));
 }
}
```
```

The ImageLinks class provides information about the image thumbnails. A null
check helps to handle cases where there's no thumbnail information available.

```
```
class ImageLinks {
 final String smallThumbnail;
 final String thumbnail;
 ImageLinks({this.smallThumbnail, this.thumbnail});
 factory ImageLinks.fromJson(Map<String, dynamic> json) {
   return ImageLinks(
       smallThumbnail: json != null? json['smallThumbnail'] :
'',
       thumbnail: json != null? json['thumbnail'] : '');
 }
}
```
```

AccessInfo CLASS

The `AccessInfo` class provides the `webReaderLink`, a link to the URL to read on the web.

```
```
class AccessInfo {
 String webReaderLink;
 AccessInfo({this.webReaderLink});
 factory AccessInfo.fromJson(Map<String, dynamic> json) {
   return AccessInfo(webReaderLink: json['webReaderLink']);
 }
}
```
```

SaleInfo CLASS

The class `SaleInfo` provides the link to buy the book as 'buyLink'.

```
```
class SaleInfo {
 final String saleability;
 final String buyLink;
 SaleInfo({this.saleability, this.buyLink});
 factory SaleInfo.fromJson(Map<String, dynamic> json) {
   return SaleInfo(saleability: json['saleability'], buyLink:
json['buyLink']);
 }
}
```
```

Now that our `BookModel` is ready to parse and create a data model for JSON data let's learn to put it in use in the next section.

CONVERTING API RESPONSE TO BookModel LIST

In this section, you'll learn to use the `BookModel` object to build a list of book entries returned from the API response. You will convert the API response to a list of `BookModel` objects. The `makeHttpCall()` function returns the `Future` of `List<BookModel>`.

```
```
//Function to make REST API call
Future<List<BookModel>> makeHttpCall() async {
 //API Key: To be replaced with your key
 final apiKey = "$YOUR_API_KEY";
 final apiEndpoint = "https://www.googleapis.com/books/v1/
volumes?key=$apiKey&q=python+coding";
 final http.Response response =
 await http.get(apiEndpoint, headers: {'Accept': 'application/
json'});
```
```

```
//Converting string response body to JSON representation
final jsonObject = json.decode(response.body);

var list = jsonObject['items'] as List;
//return the list of Book objects
return list.map((e) => BookModel.fromJson(e)).toList();
}
```

PASSING BookModel TO BookTile WIDGET

The booksListing holds the list of BookModel objects returned from the REST API call. The ListView.builder() builds the BookTile widgets for each entry passing BookModel for the current index.

```
class _BooksListingState extends State<BooksListing> {
  List<BookModel> booksListing;
  fetchBooks() async {
    var response = await makeHttpCall();
    setState(() {
      booksListing = response;
    });
  }
  @override
  Widget build(BuildContext context) {
    return Scaffold(
      body: ListView.builder(
        itemCount: booksListing == null? 0 : booksListing.
length,
        itemBuilder: (context, index) {
          //Passing bookModelObj to BookTile widget
          return BookTile(bookModelObj: booksListing[index]);
        },
      ),
    );
  }
}
```

DISPLAYING DATA

Now, data values can be accessed from the BookModel object using its members.

For example, book['volumeInfo']['title']can be accessed as bookModelObj.volumeInfo.title, and so on.

RUN THE CODE

Let's put what we have learned so far together and run the code on all four platforms: Android, iOS, MacOS, and Chrome.

Note: Don't forget to update the `apikey` with your own key. Replace "YOUR_API_KEY" with the key you obtained earlier.

FINISHED CODE (PART 2): `BookTile` WIDGET

The `BookTile` is a stateless widget that is available in *part2/booktile.dart* (Chapter 13: BookTile (Part 2)) file at GitHub repo. The `BookModel` (Chapter 13: BookModel (Part2)) is available at GitHub as well.

```
`` `

import 'package:flutter/material.dart';
import 'book.dart';

class BookTile extends StatelessWidget {
 final BookModel bookModelObj;
 const BookTile({Key key, this.bookModelObj}) : super(key:
key);
 @override
 Widget build(BuildContext context) {
   return Card(
     shape: RoundedRectangleBorder(
       borderRadius: BorderRadius.circular(10.0),
     ),
     elevation: 5,
     margin: EdgeInsets.all(10),
     child: Padding(
       padding: const EdgeInsets.all(8.0),
       child: Row(
         mainAxisAlignment: MainAxisAlignment.spaceBetween,
         children: [
           Flexible(
             child: Column(
               crossAxisAlignment: CrossAxisAlignment.start,
               children: <Widget>[
                 Text(
                   '${bookModelObj.volumeInfo.title}',
                   style: TextStyle(fontSize: 14, fontWeight:
FontWeight.bold),
                 ),
                 bookModelObj.volumeInfo.authors != null
                     ? Text(
                         'Author(s): ${bookModelObj.
volumeInfo.authors.join(", ")}',
                         style: TextStyle(fontSize: 14),
                     )
                     : Text(""),
               ],
             ),
           ),
         ),
```

```
                    bookModelObj.volumeInfo.imageLinks.thumbnail != null
                ? Image.network(
                    bookModelObj.volumeInfo.imageLinks.
thumbnail,
                        fit: BoxFit.fill,
                    )
                    : Container(),
            ],
          ),
        ),
      );
  }
}
```

FINISHED CODE (PART 2): Main METHOD

This code is available in the GitHub repo in part2 (Chapter 13: Data Modeling (main method - Part 2)) folder.

```
//importing the Dart package
import 'dart:convert';

import 'package:flutter/material.dart';
import 'package:http/http.dart' as http;

import '../../config.dart';
import 'book.dart';
import 'booktile.dart';

//Showing book listing in ListView
class BooksApp extends StatelessWidget {
 @override
 Widget build(BuildContext context) {
    return MaterialApp(
      debugShowCheckedModeBanner: false,
      home: BooksListing(),
    );
 }
}

//Making HTTP request
//Function to make REST API call
Future<List<BookModel>> makeHttpCall() async {
 //API Key: To be replaced with your key
 final apiKey = "$YOUR_API_KEY";
 final apiEndpoint =  "https://www.googleapis.com/books/v1/vol
umes?key=$apiKey&q=python+coding";
 final http.Response response =
```

```
 await http.get(apiEndpoint, headers: {'Accept': 'application/
json'});

 //Parsing API's HttpResponse to JSON format
 //Converting string response body to JSON representation
 final jsonObject = json.decode(response.body);              .

 var list = jsonObject['items'] as List;
 //return the list of Book objects
 return list.map((e) => BookModel.fromJson(e)).toList();
}

class BooksListing extends StatefulWidget {
 @override
 _BooksListingState createState() => _BooksListingState();
}

class _BooksListingState extends State<BooksListing> {
 List<BookModel> booksListing;
 fetchBooks() async {
   var response = await makeHttpCall();

   setState(() {
     booksListing = response;
   });
 }

 @override
 void initState() {
   super.initState();
   fetchBooks();
 }

 @override
 Widget build(BuildContext context) {
   return Scaffold(
     appBar: AppBar(
       title: Text("Books Listing"),
     ),
     body: ListView.builder(
       itemCount: booksListing == null?  0 : booksListing.
length,
       itemBuilder: (context, index) {
         //Passing bookModelObj to BookTile widget
         return BookTile(bookModelObj: booksListing[index]);
       },
     ),
   );
 }
}
```

FIGURE 13.5 Final code run results for Android platform

ANDROID TARGET

The following screenshot (Figure 13.5) is taken on the *Pixel 3 API 28* emulator.

iOS TARGET

The following screenshot (Figure 13.6) is taken from iPhone SE (2nd generation) simulator. This is the default simulator selected for my XCode configuration.

FIGURE 13.6   Final code run results for iOS platform

DESKTOP (macOS)

Figure 13.7 shows the book listing in a desktop app running at macOS platform.

WEB (CHROME)

Figure 13.8 shows the book listing for web in Chrome browser.

SOURCE CODE ONLINE

The source code for this chapter (Chapter 13) is available online at GitHub.

FIGURE 13.7   Final code run results for macOS platform

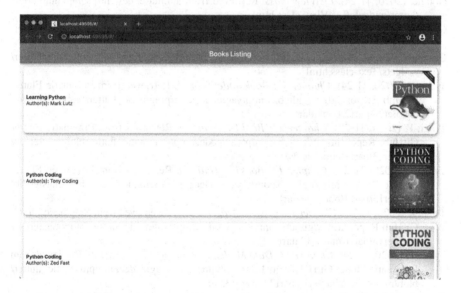

FIGURE 13.8   Final code run results for Chrome platform

CONCLUSION

Congratulations! You've made your own book listing app using Flutter. In this chapter, you learned how to parse API responses in JSON and access needed book information for the app. We learned to parse HttpResponse fetched from API into JSON format. Finally, you learned to map JSON responses in data model classes.

REFERENCES

Dart Dev. (2020, 12 22). *LinkedHashMap<K, V> class*. Retrieved from Dart API: https://api. dart.dev/stable/2.8.4/dart-collection/LinkedHashMap-class.html

Dart Team. (2020, 11 24). *dart:convert library*. Retrieved from api.dart.dev: https://api.dart. dev/stable/2.7.1/dart-convert/dart-convert-library.html

Dart Team. (2020, 11 24). *HttpResponse class*. Retrieved from api.dart.dev: https://api.dart. dev/stable/2.7.1/dart-io/HttpResponse-class.html

Dart Team. (2020, 11 24). *Map<K, V> class*. Retrieved from api.dart.dev: https://api.dart.dev/ stable/2.8.4/dart-core/Map-class.html

Flutter Team. (2020, 11 24). *BuildContext class*. Retrieved from api.flutter.dev: https://api. flutter.dev/flutter/widgets/BuildContext-class.html

Flutter Team. (2020, 11 24). *ListView class*. Retrieved from api.dart.dev: https://api.flutter. dev/flutter/widgets/ListView-class.html

Google. (2020, 11 24). *Card class*. Retrieved from api.flutter.dev: https://api.flutter.dev/flutter/ material/Card-class.html

Google. (2020, 11 24). *Cards*. Retrieved from material.io: https://material.io/components/ cards

Google. (2020, 11 24). *Column class*. Retrieved from api.flutter.dev: https://api.flutter.dev/ flutter/widgets/Column-class.html

Google. (2020, 11 24). *Flexible class*. Retrieved from api.flutter.dev: https://api.flutter.dev/ flutter/widgets/Flexible-class.html

Google. (2020, 11 24). *ListView class*. Retrieved from api.flutter.dev: https://api.flutter.dev/ flutter/widgets/ListView-class.html

Google. (2020, 11 24). *Row class*. Retrieved from api.flutter.dev: https://api.flutter.dev/flutter/ widgets/Row-class.html

Google. (2020, 11 24). *Text class*. Retrieved from api.flutter.dev: https://api.flutter.dev/flutter/ widgets/Text-class.html

Tyagi, P. (2020, 11 24). *Chapter 13: BookModel (Part2)*. Retrieved from Pragmatic Flutter GitHub Repo: https://github.com/ptyagicodecamp/pragmatic_flutter/blob/master/lib/ chapter13/part2/book.dart

Tyagi, P. (2020, 11 24). *Chapter 13: BookTile (Part 1)*. Retrieved from Pragmatic Flutter GitHub Repo: https://github.com/ptyagicodecamp/pragmatic_flutter/blob/master/lib/ chapter13/part1/booktile.dart

Tyagi, P. (2020, 11 24). *Chapter 13: BookTile (Part 2)*. Retrieved from Pragmatic Flutter GitHub Repo: https://github.com/ptyagicodecamp/pragmatic_flutter/blob/master/lib/ chapter13/part2/booktile.dart

Tyagi, P. (2020, 11 24). *Chapter 13: Custom Widget (Part 1)*. Retrieved from Pragmatic Flutter GitHub Repo: https://github.com/ptyagicodecamp/pragmatic_flutter/blob/master/lib/ chapter13/part1/main_13.dart

Tyagi, P. (2020, 11 24). *Chapter 13: Data Modeling (main method - Part 2)*. Retrieved from Pragmatic Flutter GitHub Repo: https://github.com/ptyagicodecamp/pragmatic_flutter/ blob/master/lib/chapter13/part2/main_13.dart

Tyagi, P. (2020, 12 22). *Chapter 13*. Retrieved from Pragmatic Flutter GitHub Repo: https:// github.com/ptyagicodecamp/pragmatic_flutter/tree/master/lib/chapter13

Tyagi, P. (2021). Chapter 12: Integrating REST API. In P. Tyagi, *Pragmatic Flutter: Building Cross-Platform Mobile Apps for Android, iOS, Web & Desktop*. CRC Press.

14 Navigation and Routing

In the previous chapter (Chapter 12: Integrating REST API), we learned to make RESTful HTTP requests to Books API and fetch book listings data. In last chapter (Chapter 13: Data Modeling), the book listing was rendered in the `ListView` widget using data model class. A custom widget – `BookTile` was used to render each list item. The `BookTile` widget displayed only three pieces of information about the book: title, author(s), and image of the cover page. The `ListView` widget displayed a `BookTile` widget for each item returned from API. Next, we want to show additional details about the book like its publication date, publisher, description, etc. Each book has its own detailed information page `BookDetailsPage`. In this chapter, we will start creating this simple page – `BookDetailsPage`, by rendering the book's description. Later on, we will learn to build navigation from the `BookListing` screen to the `BookDetailsPage` screen for selected list item in the `BookListing` screen. Figure 14.1 shows the homepage for the *BooksApp* that we created in the previous chapter (Chapter 13: Data Modeling).

In this chapter, we will learn about the three types of navigation and routing to implement navigation from the `BookListing` homepage to the `BookDetailsPage` screen. The `BookDetailsPage` page only displays the book details as shown in Figure 14.2.

In the next chapter (Chapter 15: The Second Page: BookDetails), we will continue building the `BookDetailsPage` interface.

SIMPLE `BookDetailsPage` SCREEN

In this section, you will learn to build a basic secondary page widget – `BookDetailsPage`, which is used as a placeholder to understand navigation and routing concepts in Flutter application. The anatomy of this simple page looks like as shown in Figure 14.3.

The `BookDetailsPage` is a `StatelessWidget` with two children widgets: `AppBar` and `Center`. The `AppBar` widget displays the book's title, and the `Center` widget is assigned to the `body` property for the `Scaffold` widget. This `Center` widget has a child `Text` widget to display the book's detailed description.

```
```
import 'package:flutter/material.dart';
import 'book.dart';
class BookDetailsPage extends StatelessWidget {
  final BookModel book;
  const BookDetailsPage({Key key, this.book}) : super(key: key);
  @override
```
```

FIGURE 14.1 BookListing – The homepage of BooksApp. Listing books using the BookTile widget

```
Widget build(BuildContext context) {
  return Scaffold(
    appBar: AppBar(
      title: Text(book.volumeInfo.title),
    ),
    body: Center(
      child: Text(book.volumeInfo.description),
    ),
  );
}
}
```

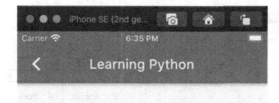

FIGURE 14.2 BookDetailsPage – Simple page for book details. Only shows title in the appBar and book's description in the main area

The data structure `BookModel` was constructed in the chapter (Chapter 13: Data Modeling) from the Books API's JSON response. The `BookDetailsPage` requires the currently selected book item information passed along to be able to render its title and description. The `book` object of `BookModel` data type provides the book's title information as `book.volumeInfo.title`, and its description is available as `book.volumeInfo.description`. At this point, we display only the title and description in the book details page. In the next chapter (Chapter 15: The Second Page: BookDetails), we will build a more detailed interface to display selected book information.

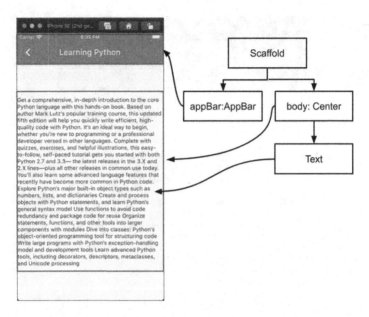

FIGURE 14.3   Anatomy of simple BookDetailsPage widget

NAVIGATOR WIDGET

The Flutter framework implements navigation across multiple pages using the `Navigator` (Navigator class) widget. It's a widget to manage children widgets using a stack discipline. There are three different ways to implement navigation in the Flutter application.

- **Direct Navigation**: The direct navigation is also known as Unnamed Routing. It is implemented with the help of `MaterialPageRoute` (MaterialPageRoute<T> class).
- **Static Navigation**: The static navigation is a type of Named Routing. It is implemented by assigning a map of routes to `MaterialApp` `routes` (routes property) property. The `routes` property acts like the application's top-level routing table. The route name is pushed using `Navigator.pushNamed(...)`. The routing table decides which route will map to what widget.
- **Dynamic Navigation**: The dynamic navigation is a type of Named Routing as well. In this navigation type, routes are generated by implementing the `onGenerateRoute` (onGenerateRoute property) callback in the `MaterialApp` class. It's a function that provides the routes dynamically. This routing function is assigned to the `onGenerateRoute` property of `MaterialApp`, as mentioned earlier. The route name is pushed using `Navigator.pushNamed(...)` similar to static navigation.

Let's explore these three types of routings next.

DIRECT NAVIGATION

Direct navigation is also known as unnamed routing. As mentioned earlier, it is implemented using `MaterialPageRoute`. The `MaterialPageRoutes` is pushed directly to the navigator widget stack. This approach can contribute to duplicate and boilerplate code across multiple pages. This boilerplate code multiplies with growing screens/pages. It is challenging to keep track of logic wrapped around these routes in a commonplace since it spreads around multiple classes.

Let's implement this type of navigation in our *BooksApp* for navigating from the `BookListing` page to `BookDetailsPage`.

ENTRY POINT

The `MaterialApp` assigns the `BooksListing` screen to its home property.

```
class BooksApp extends StatelessWidget {
 @override
 Widget build(BuildContext context) {
   //Using Direct Navigation (un-named routing)
   return MaterialApp(
     debugShowCheckedModeBanner: false,
     home: BooksListing(),
   );
 }
}
```

NAVIGATION IMPLEMENTATION

This routing is implemented using `Navigator.push()` (push<T extends Object> method) method. The `MaterialPageRoute` (MaterialPageRoute<T> class) is pushed on the `Navigator` (Navigator class). The Navigator is a widget that manages a set of child widgets as a stack. These child widgets are pages or screens pushed on the Navigator widget. The Navigator widget refers to these children as `Route` (Route class) objects.

DETECTING GESTURE

The navigation is initiated from the user activity on the homepage screen. That means the `BookListing` page list items need to be interacted with to navigate to its detailed page. The `GestureDetector` (GestureDetector class) widget is used to detect the gestures. It handles the taps on the listing items using its `onTap:` property.

```
class BooksListing extends StatefulWidget {
 @override
 _BooksListingState createState() => _BooksListingState();
```

```
}

class _BooksListingState extends State<BooksListing> {
 List<BookModel> booksListing;
 ...
 @override
 Widget build(BuildContext context) {
   return Scaffold(
     ...
     body: ListView.builder(
       itemCount: booksListing == null?  0 : booksListing.
length,
       itemBuilder: (context, index) {
         //Passing bookModelObj to BookTile widget
         return GestureDetector(
           child: BookTile(bookModelObj: booksListing[index]),
           onTap: () {});
       },
     ),
   );
 }
}
```

PASSING DATA

The `MaterialPageRoute` uses a builder to build the primary contents of the route
(page/screen). The `book` object is passed as an argument to the `BookDetailsPage`
widget.

```
onTap: () {
 Navigator.of(context).push(
   MaterialPageRoute(
     builder: (context) => BookDetailsPage(
       book: booksListing[index],
     ),
   ),
 );
},
```

TIP

The above code can also be written in one line without using curly braces.

```
onTap: () =>
 Navigator.of(context).push(
   MaterialPageRoute(
```

```
    builder: (context) => BookDetailsPage(
      book: booksListing[index],
    ),
  ),
)
~ ~ ~
```

SOURCE CODE

The full source code of this example (Chapter 14: Direct Navigation) is available on
GitHub.

STATIC NAVIGATION

In a static navigation application's top-level routing table is implemented using a
Map (Map<K, V> class) of routes (pages/screens). This routing table is assigned
to MaterialApp routes (routes property) property. The route names for pages
are pushed using Navigator.pushNamed(...) (pushNamed<T extends Object>
method). This routing is known as Named Routing because each page is given a
unique name, which is pushed on the Navigator widget.

The MaterialApp and WidgetApp provide the routes property. This property enables assigning routes as Map<String, WidgetBuilder>. This option
works better when there are not many logical steps wrapped around the routes. For
example, authentication or verification related code wrapped around the logic for
navigating to a particular page. In this type of navigation, only the app's global data
can be passed on to the second page.

ENTRY POINT

The static navigation provides two ways to assign the initial page – BooksListing
widget. One option is to assign BooksListing widget to the home property.
Second option is to assign a route Map containing </, BooksListing()> entry
to routes property. The '/' stands for the home page mapping.

```
~ ~ ~
//The booksListing data is available global to app
List<BookModel> booksListing;
class BooksApp extends StatelessWidget {
 @override
 Widget build(BuildContext context) {
   //Using Static Navigation (Named Routing)
   return MaterialApp(
     debugShowCheckedModeBanner: false,
     //home: BooksListing(),
     //Named-Routing using Map routing-table
     routes: <String, WidgetBuilder>{
       '/': (BuildContext context) => BooksListing(),
       '/details': (BuildContext context) => BookDetailsPage(
```

```
        book: booksListing[0],
      ),
    },
  );
 }
}
~ ~ ~
```

In static navigation, the routing table is assigned statically. That means any object can be passed in the routing map only. This requires data to be available at the top-level. We need to pass the `book` object to the `BookDetailsPage` for the `/details` route name. As you can see that this value is static and cannot be changed with the update in the book selection in the `BookListing` book items list. The value is retrieved by accessing `List<BookModel> booksListing`, which gets updated after the REST API response is returned. For the `/details` route, only one book object is assigned for the lifecycle of the app as `book: booksListing[0]`. This causes to show the same book's details for every book displayed in the `BookListing` widget.

NAVIGATION IMPLEMENTATION

All routes/pages have entries in the routing table above. The Map entry '</details, BookDetailsPage()>'is added to navigate to the `BookDetailsPage` screen. The '/details' is the alias/name to the `BookDetailsPage` screen. This name is pushed on the Navigator widget using `Navigator.pushNamed` (pushNamed<T extends Object> method).

DETECTING GESTURE

The navigation is initiated from the user activity on the homepage screen, similar to a direct navigation example. The `GestureDetector` widget is used to detect the gestures. It handles the tap gesture on the listing with 'onTap: ' property. Let's check out the new way of pushing routes on the `Navigator` widget in the code below. Please note the named route '/details' are pushed on the Navigator stack. This route is declared in the `MaterialApp` routing table.

```
~ ~ ~
onTap: () =>
    Navigator.pushNamed(context, '/details')
~ ~ ~
```

PASSING DATA

As we saw earlier, the data can be passed to the `BookDetailsPage()` at the top-level only when the routes are assigned to `routes` property. In this case, only globally available data can be passed to another widget. In this implementation, only the first item `booksListing[0]` detail page is available for any selection on the homepage.

Source Code

The full source code of this example (Chapter 14: Static Navigation) is available on GitHub.

DYNAMIC NAVIGATION

In dynamic navigation, routes are generated dynamically with the help of a function. This function implements the onGenerateRoute (onGenerateRoute property) callback in the MaterialApp class. This is a type of Named Routing that makes use of onGenerateRoute property.

The MaterialApp and WidgetApp provide the onGenerateRoute property to assign the callback function, say generateRoute returning a route. It allows the data to pass using RouteSettings (RouteSettings class). It carries the data to help construct a Route (Route class).

Any authorization or verification logic can be extracted to a single place. This routing provides the option to show a default page when a route or match is not found. In our *BooksApp*, we will use the `PageNotFound` widget when no route is matched. It's a simple page that displays the message that the requested page is not available.

Entry Point

The entry page BookListing is assigned to home property. The initialRoute property can be used to set the beginning route/page. The generateRoute callback function handles the navigational logic.

```
class BooksApp extends StatelessWidget {
  @override
  Widget build(BuildContext context) {
    //Using Dynamic Navigation (Named Routing)
    return MaterialApp(
      debugShowCheckedModeBanner: false,
      home: BooksListing(),
      //Named with onGenerateRoute
      initialRoute: '/',
      onGenerateRoute: generateRoute,
    );
  }
}
```

THE generateRoute() FUNCTION

The generateRoute() function takes the RouteSettings as an argument, which allows sending data along. The arguments property on the RouteSettings object retrieves any arguments sent with the widget.

```
` ` `
Route<dynamic> generateRoute(RouteSettings routeSettings) {
 final args = routeSettings.arguments;
 switch (routeSettings.name) {
   case '/':
     return MaterialPageRoute(
       builder: (context) => BooksListing(),
     );

   case '/details':
     if (args is BookModel) {
       return MaterialPageRoute(
         builder: (context) => BookDetailsPage(
           book: args,
         ),
       );
     }

     return MaterialPageRoute(
       builder: (context) => PageNotFound(),
     );

   default:
     return MaterialPageRoute(
       builder: (context) => PageNotFound(),
     );
 }
}
` ` `
```

NAVIGATION IMPLEMENTATION

The Navigator uses the Route object to represent the page/screen. The genera-teRoute() function returns the appropriate route based on the matching name. The RouteSettings is useful in passing around these route names and arguments, if any. The route name is extracted using routeSettings.name. The arguments can be extracted using routeSettings.arguments. When no match is found, a common default page is shown to display the appropriate message.

DETECTING GESTURE

The navigation is initiated from the user activity on the homepage screen, similar to direct and static navigation examples. The GestureDetector widget is used to detect the gestures. It handles the tap gesture on the listing with `onTap:` property.

PASSING DATA

Dynamic navigation uses named-routing as well. This routing allows passing selected book object `booksListing[index]` as an argument to generateRoute()

callback function. The generateRoute() function extracts the route name and its arguments using the RouteSettings object. Refer to generateRoute() to understand extracting route names and arguments from the RouteSettings object.

```
onTap: () =>
    Navigator.pushNamed(
        context,
        '/details',
        arguments: booksListing[index],
    )
```

SOURCE CODE

The full source code of this example (Chapter 14: Dynamic Navigation) is available on GitHub.

CONCLUSION

In this chapter, we learned to navigate from one page to another. We created a simple second page to navigate from the first page. In the next chapter, we will work on the second page to layout book's details more intuitively.

REFERENCES

Chapter 14: Dynamic Navigation. (2020, 11 25). Retrieved from Pragmatic Flutter GitHub Repo: https://github.com/ptyagicodecamp/pragmatic_flutter/blob/master/lib/chapter14/main_14_dynamic.dart

Dart Team. (2020, 11 25). *Map<K, V> class*. Retrieved from api.dart.dev: https://api.dart.dev/stable/2.8.4/dart-core/Map-class.html

Flutter Team. (2020, 11 25). *MaterialPageRoute<T> class*. Retrieved from api.flutter.dev: https://api.flutter.dev/flutter/material/MaterialPageRoute-class.html

Flutter Team. (2020, 11 25). *pushNamed<T extends Object> method*. Retrieved from Flutter Dev: https://api.flutter.dev/flutter/widgets/Navigator/pushNamed.html

Flutter Team. (2020, 11 25). *routes property*. Retrieved from api.flutter.dev: https://api.flutter.dev/flutter/material/MaterialApp/routes.html

GestureDetector class. (2020, 11 24). Retrieved from Flutter Dev: https://api.flutter.dev/flutter/widgets/GestureDetector-class.html

Google. (2020, 11 25). *Navigator class*. Retrieved from api.flutter.dev: https://api.flutter.dev/flutter/widgets/Navigator-class.html

Google. (2020, 11 25). *onGenerateRoute property*. Retrieved from api.flutter.dev: https://api.flutter.dev/flutter/widgets/Navigator/onGenerateRoute.html

Google. (2020, 11 25). *onGenerateRoute property*. Retrieved from Flutter Dev: https://api.flutter.dev/flutter/material/MaterialApp/onGenerateRoute.html

Google. (2020, 11 25). *push<T extends Object> method*. Retrieved from Flutter Dev: https://api.flutter.dev/flutter/widgets/Navigator/push.html

Google. (2020, 11 25). *Route class*. Retrieved from Flutter Dev: https://api.flutter.dev/flutter/widgets/Route-class.html

Google. (2020, 11 25). *RouteSettings class*. Retrieved from Flutter Dev: https://api.flutter.dev/
flutter/widgets/RouteSettings-class.html

Tyagi, P. (2020, 11 25). *Chapter 14: Direct Navigation*. Retrieved from Pragmatic Flutter
GitHub Repo: https://github.com/ptyagicodecamp/pragmatic_flutter/blob/master/lib/
chapter14/main_14_direct.dart

Tyagi, P. (2020, 11 25). *Chapter 14: Static Navigation*. Retrieved from Pragmatic Flutter
GitHub Repo: https://github.com/ptyagicodecamp/pragmatic_flutter/blob/master/lib/
chapter14/main_14_static.dart

Tyagi, P. (2021). Chapter 12: Integrating REST API. In P. Tyagi, *Pragmatic Flutter: Building
Cross-Platform Mobile Apps for Android, iOS, Web & Desktop*. CRC Press.

Tyagi, P. (2021). Chapter 15: The Second Page: BookDetails. In P. Tyagi, *Pragmatic Flutter:
Building Cross-Platform Mobile Apps for Android, iOS, Web & Desktop*. CRC Press.

Tyagi, P. (n.d.). Chapter 13: Data Modeling. In P. Tyagi, *Pragmatic Flutter: Building Cross-
Platform Mobile Apps for Android, iOS, Web & Desktop*. CRC Press.

15 The Second Page – BookDetailsPage Widget

In previous chapter (Chapter 12: Integrating REST API), we learned to make REST calls to Google Books API and fetch book listings. In chapter (Chapter 13: Data Modeling), we learned to render book listings in a list using the `ListView` widget. We also learned to create a custom widget – `BookTile` to generate each list item. The `BookTile` widget displayed only three essential information about the book: title, author(s), and image of the cover page. Later in chapter (Chapter 14: Navigation & Routing), we learned about navigation from the homepage to another page. We started working on adding the second page – `BookDetailsPage` to display detailed information about the selected book. However, the `BookDetailsPage` only showed the book's title in the `appBar` and its description in the screen's main area.

In this chapter, we will continue designing and implementing `BookDetailsPage` to display more relevant information about the given book.

ANATOMY OF `BookDetailsPage` WIDGET

Let's start listing out the pieces of information that we are interested in displaying about the book.

- Book's title and subtitle.
- Author(s) listing.
- Publication details like publisher and publishing date.
- Cover page imagery.
- Web reader link to preview book sample.
- Link to buy the book, if it is available for sale.
- Detailed book description.

Typically, the above can be categorized into three groups: quick book information, action items, and detailed description. Let's rearrange the above information per their groups.

1. **Book Information**: Title, subtitle, authors, publisher, publication date, and cover imagery.
2. **Action items**: Web reader and buying links.
3. **Detailed Description**: Book's detailed description.

Figure 15.1 shows the labeled interface for BookDetailsPage.

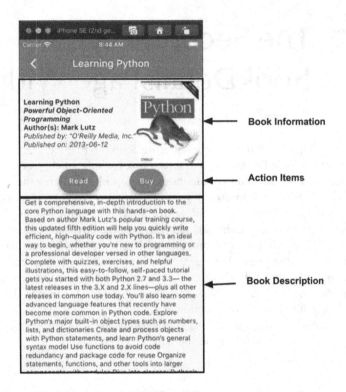

Book Information

Action Items

Book Description

FIGURE 15.1   Labeled interface for BookDetailsPage

`BookDetailsPage` SCREEN'S LAYOUT

Let's layout the above three main interface elements in the `BookDetailsPage` screen, as shown in Figure 15.2. Like any other Flutter page, the `Scaffold` widget is the root of the page. The `Scaffold` widget has an `AppBar` widget to display the book's title and the arrow button to navigate back to the last screen. The `body` property is assigned to `SingleChildScrollView` (SingleChildScrollView class). It provides a box in which a widget can be scrolled. The scrolling widget is used to avoid the content overflow problem that may surface when the book's description is too long to fit on the screen. The `SingleChildScrollView` widget occupies all the space on the screen and doesn't leave any spacing around it. If we put any widget inside, it will expand to full screen as well. We will use the `Padding` widget to inset its child by `8.0` around all sides. Next, we want to vertically align book information, actions, and book description data. The `Column` widget helps to align its children in a vertical array. The `Column` widget uses the following three custom widgets to organize the book's information on the `BookDetailsPage` screen.

1. InformationWidget
2. ActionsWidget
3. DescriptionWidget

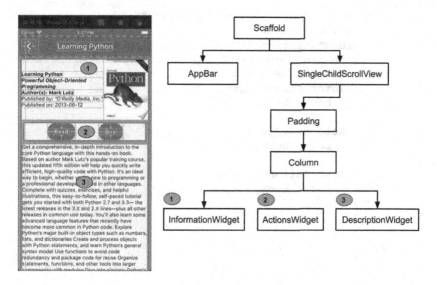

FIGURE 15.2　Anatomy of BookDetailsPage

Refer to Figure 15.2 to understand the mapping visually.

IMPLEMENTING `BookDetailsPage` WIDGET

Now that you have an understanding of how the widgets are arranged in
BookDetailsPage, it's time to dive into code. The BookDetailsPage is a
StatelessWidget. It receives the BookModel object `book` passed from the
BookListing page. The title is retrieved as `book.volumeInfo.title`, and displayed in the AppBar. The Column widget's contents are aligned to the starting off the
screen using `crossAxisAlignment: CrossAxisAlignment.start`. The
extra vertical space is divided evenly between children using `mainAxisAlign-
ment: MainAxisAlignment.spaceBetween`. The Column widget has three
children: InformationWidget, ActionsWidget, and DescriptionWidget.
All three widgets are passed the BookModel object to render relevant information.

```
```
class BookDetailsPage extends StatelessWidget {
  final BookModel book;
  const BookDetailsPage({Key key, this.book}) : super(key:
key);
  @override
  Widget build(BuildContext context) {
    return Scaffold(
      appBar: AppBar(
        title: Text(book.volumeInfo.title),
      ),
      body: SingleChildScrollView(
        child: Padding(
```
```

```
        padding: const EdgeInsets.all(8.0),
        child: Column(
          crossAxisAlignment: CrossAxisAlignment.start,
          mainAxisAlignment: MainAxisAlignment.spaceBetween,
          children: <Widget>[
            InformationWidget(
              book: book,
            ),
            ActionsWidget(
              book: book,
            ),
            DescriptionWidget(
              book: book,
            ),
          ],
        ),
      ),
     ),
    ),
   );
  }
}
```

InformationWidget

The InformationWidget builds the top part of the BookDetailsPage. It displays the book's title, subtitle, authors, publisher, publishing date, and cover imagery. There's a total six-piece of information that needs to be displayed. The first five elements stay to the left of the screen, and cover imagery is aligned to the right side of the screen. All elements are aligned horizontally. It makes sense to use the Row widget as the parent widget since it aligns its children in a horizontal array. The five elements to the right are arranged in a Flexible widget. The Column widget is assigned as a child to the Flexible widget. The Flexible widget provides the flexibility to its Column child to expand to fill the available space vertically. The Column holds all the five Text widgets as its children. Refer to Figure 15.3 to see the visual representation of the InformationWidget layout.

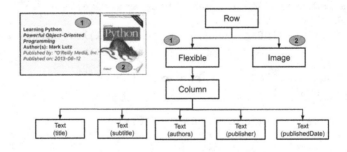

FIGURE 15.3 Anatomy of InformationWidget

Let's check out the `InformationWidget` code implementation below:

```
class InformationWidget extends StatelessWidget {
 final BookModel book;

 const InformationWidget({Key key, this.book}) : super(key:
key);

 @override
 Widget build(BuildContext context) {
   return Row(
     //Divides extra space evenly horizontally
     mainAxisAlignment: MainAxisAlignment.spaceEvenly,
     children: [
       Flexible(
         child: Column(
           //Children are aligned to start of the screen
           crossAxisAlignment: CrossAxisAlignment.start,
           //Divides extra space evenly vertically
           mainAxisAlignment: MainAxisAlignment.spaceEvenly,
           children: <Widget>[
             //Book Title
             //Book SubTitle
             //Author(s)
             //Publisher
             //PublishedDate
           ],
         ),
       ),
       //Displaying cover image
     ],
   );
 }
}
```

The book title, subtitle, author(s), publisher, and published date are rendered in a Text widget. The image is rendered in the `Image` widget. Only the non-null values are rendered in the `Text` and/or `Image` widget; otherwise, an empty `Container` widget is rendered. A style is applied to `Text` widgets to font size as fourteen and bold font weight.

```
style: TextStyle(fontSize: 14, fontWeight: FontWeight.bold)
```

Let's checkout implementations for each of the widgets in the code below:

```
class InformationWidget extends StatelessWidget {
```

```
final BookModel book;
const InformationWidget({Key key, this.book}) : super(key:
key);
@override
Widget build(BuildContext context) {
  return Row(
    mainAxisAlignment: MainAxisAlignment.spaceEvenly,
    children: [
      Flexible(
        child: Column(
          crossAxisAlignment: CrossAxisAlignment.start,
          mainAxisAlignment: MainAxisAlignment.spaceEvenly,
          children: <Widget>[
            //Book Title
            book.volumeInfo.title != null
                ? Text(
                    '${book.volumeInfo.title}',
                    style:
                        TextStyle(fontSize: 14, fontWeight:
FontWeight.bold),
                  )
                : Container(),

            //Book subtitle
            book.volumeInfo.subtitle != null
                ? Text(
                    '${book.volumeInfo.subtitle}',
                    style: TextStyle(
                        fontSize: 14,
                        fontWeight: FontWeight.bold,
                        fontStyle: FontStyle.italic),
                  )
                : Container(),
            //Book authors. Used join() method on list to
convert list into comma-separated String
            book.volumeInfo.authors != null
                ? Text(
                    'Author(s): ${book.volumeInfo.authors.
join(", ")}',
                    style:
                        TextStyle(fontSize: 14, fontWeight:
FontWeight.bold),
                  )
                : Container(),

            //Publisher
            book.volumeInfo.publisher != null
                ? Text(
                    "Published by: ${book.volumeInfo.
publisher}",
                    style:
```

```
                    TextStyle(fontSize: 14, fontStyle:
FontStyle.italic),
                )
            : Container(),
          //PublishedDate
          book.volumeInfo.publishedDate != null
            ? Text(
                "Published on: ${book.volumeInfo.
publishedDate}",
                style:
                    TextStyle(fontSize: 14, fontStyle:
FontStyle.italic),
                )
            : Container(),
        ],
      ),
    ),
    //Rendering cover page image
    book.volumeInfo.imageLinks.thumbnail != null
      ? Image.network(
          book.volumeInfo.imageLinks.thumbnail,
          fit: BoxFit.fill,
          )
      : Container(),
    ],
  );
}
}
```

ActionsWidget

The ActionsWidget renders two material design floating action buttons (FABs) to let users read the sample and/or buy the book by clicking on respective buttons. These buttons are implemented using the FloatingActionButton (FloatingActionButton class) widget. Clicking on each button, launch the URL provides by Books API for the given book. For launching URL, the url _ launcher plugin (url_launcher plugin) is used. At the time of this writing, the current version for url _ launcher is '5.4.11'. Add this dependency in *pubspec.yaml* under the `dependencies` section.

```
dependencies:
 #To open web reader link and buyLink from BookDetailsPage
 url_launcher: ^5.4.11
```

The ActionsWidget sits below the InformationWidget in BookDetailsPage. The ActionsWidget has a Padding widget at the top. The Padding widget has a child Row widget. The Row widget aligns its children horizontally. This Row widget

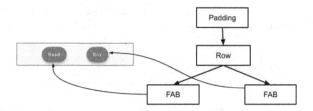

FIGURE 15.4 Anatomy of ActionsWidget

has two children of type `FloatingActionButton` or FAB with the extended variant. The `FloatingActionButton.extended()` method allows FAB to be more giant and have icons and label as well. The `onPressed:` has the implementation to launch the URL for FAB. Refer to Figure 15.4 to see the visual representation of the `ActionsWidget` layout.

Let's check out the `ActionsWidget` code implementation:

```
```
class ActionsWidget extends StatelessWidget {
 final BookModel book;
 const ActionsWidget({Key key, this.book}) : super(key: key);
 @override
 Widget build(BuildContext context) {
   return Padding(
     padding: const EdgeInsets.all(8.0),
     child: Row(
       mainAxisAlignment: MainAxisAlignment.spaceEvenly,
       children: [
         book.accessInfo.webReaderLink != null
             ? FloatingActionButton.extended(
                 label: Text("Read"),
                 heroTag: "webReaderLink",
                 onPressed: () => launch(book.accessInfo.
webReaderLink),
)
             : Container(),
         book.saleInfo.saleability == "FOR_SALE"
             ? FloatingActionButton.extended(
                 label: Text("Buy"),
                 heroTag: "buy_book",
                 onPressed: () => launch(book.saleInfo.
buyLink),
)
             : Container(),
       ],
     ),
   );
 }
}
```
```

DescriptionWidget

The description widget is a simple widget to render the book's description in the `Text` widget. This widget is not required to be in its own widget. However, I extracted this widget on its own to keep things simple. Here's the `DescriptionWidget` implementation.

```
class DescriptionWidget extends StatelessWidget {
  final BookModel book;
  const DescriptionWidget({Key key, this.book}) : super(key: key);
  @override
  Widget build(BuildContext context) {
    return book.volumeInfo.description != null
        ? Text(book.volumeInfo.description.toString())
        : Container();
  }
}
```

SOURCE CODE ONLINE

The full source code for this example (Chapter 15: BookDetailsPage) is available on GitHub.

CONCLUSION

In this chapter, we implemented the `BookDetailsPage` screen to display detailed book information. The page's body shows the necessary information like title, author(s), and publication details in the top part. The middle part has the action items to preview the book sample and purchase the book. The bottom part displays the detailed book description.

REFERENCES

Flutter Team. (2020, 11 25). *url_launcher plugin*. Retrieved from pub.dev: https://pub.dev/packages/url_launcher

Google. (2021, 11 25). *FloatingActionButton class*. Retrieved from api.flutter.dev: https://api.flutter.dev/flutter/material/FloatingActionButton-class.html

Google. (2021, 11 25). *SingleChildScrollView class*. Retrieved from api.flutter.dev: https://api.flutter.dev/flutter/widgets/SingleChildScrollView-class.html

Tyagi, P. (2020, 11 25). *Chapter 15: BookDetailsPage*. Retrieved from Pragmatic Flutter GitHub Repo: https://github.com/ptyagicodecamp/pragmatic_flutter/blob/master/lib/chapter15/book_details_page.dart

Tyagi, P. (2021). Chapter 12: Integrating REST API. In P. Tyagi, *Pragmatic Flutter: Building Cross-Platform Mobile Apps for Android, iOS, Web & Desktop*. CRC Press.

Tyagi, P. (2021). Chapter 13: Data Modeling. In P. Tyagi, *Pragmatic Flutter: Building Cross-Platform Mobile Apps for Android, iOS, Web & Desktop*. CRC Press.

Tyagi, P. (2021). Chapter 14: Navigation & Routing. In P. Tyagi, *Pragmatic Flutter: Building Cross-Platform Mobile Apps for Android, iOS, Web & Desktop*. CRC Press.

16 Introduction to State Management

Any real-world application needs to handle the application state sooner or later. As developers, we might start to build an application with one screen, but soon multiple screens supporting different features join the band. These additional screens may need to know the state of the application at a given time. For example, for an e-commerce shopping application, the cart page should be updated based on what was selected from the catalog on the page before it.

When it comes to building Flutter applications where everything is a 'widget', the deeply nested widget trees start to build up quickly. The widgets in widget-trees might need to share the application's state and pass their state to other widgets. It becomes crucial to handle the widget's state sharing or 'State Management' appropriately and efficiently to avoid the scaling issues that can lead to technical debts.

In Flutter development, the architecture patterns and state management are used interchangeably. However, architecture patterns help to streamline code organization into separate layers to segregate responsibilities. There are many options to manage the state of the application in the Flutter applications. The Flutter app states are categorized into two types: Local (or Ephemeral) State and Application State. In broader terms, the local state refers to the state scoped to a single widget, whereas the application state is application-wide. Refer to Flutter's official documentation (Differentiate between ephemeral state and app state) to learn about state management in detail.

The local state can be managed using a combination of `StatefulWidget` (StatefulWidget class) and `setState()` (setState method) methods. However, there are several solutions available for managing an application-wide state. We'll use Flutter's default sample *CounterApp* to understand the various state-management solutions throughout several next chapters. I have preferred this example to show the contrast in state-management implementations while, still, full code can fit in a single page.

REVISITING DEFAULT *CounterApp*

Flutter creates a simple counter app by default when a Flutter project is created either by command line or Android studio. The following command creates the default counter app.

```
flutter create.
```

It is a basic app with an app bar, two `Text` widgets in the middle of the screen to display the counter information, and a `FloatingActionButton` (or FAB) widget

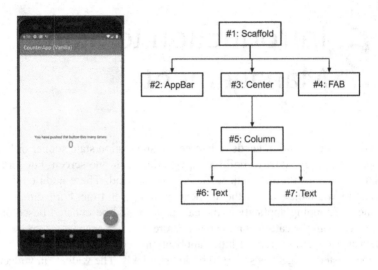

FIGURE 16.1   Anatomy of CounterApp for vanilla implementation

to increase the counter every time it clicked. There is a total of seven widgets in this app:

1. Scaffold: The root widget.
2. AppBar: Displays title of the app.
3. Center: Aligns its children in the center of the screen.
4. FloatingActionButton: FAB to increase the counter.
5. Column: Align its children vertically.
6. Text: Displays the information text.
7. Text: Displays the updated counter number.

Refer to Figure 16.1 for the visual structure of the app's widgets.

VANILLA PATTERN

This sample app is a classic example showcasing local state management. This app comes with local state management in-built. Using different state-management approaches, we will be using this example to increase the counter and display its value. The goal is to introduce various state-management solutions and understand the implementation differences between different approaches while achieving the same purpose. In this chapter, we'll be focusing on basic or vanilla implementation without any whistles and bells.

Let's get started with the default in-built state-management solution that comes with the *CounterApp*. It uses StatefulWidget in combination with the `set-State()` method to update and reflect a widget state at a given time. Let's check out the code for the app below:

```
```

import 'package:flutter/material.dart';

void main() {
 runApp(CounterApp());
}

class CounterApp extends StatelessWidget {
 // This widget is the root of your application.
 @override
 Widget build(BuildContext context) {
   return MaterialApp(
     debugShowCheckedModeBanner: false,
     theme: ThemeData(
       primarySwatch: Colors.blue,
       visualDensity: VisualDensity.adaptivePlatformDensity,
     ),
     home: MyHomePage(title: 'CounterApp (Vanilla)'),
   );
 }
}

class MyHomePage extends StatefulWidget {
 MyHomePage({Key key, this.title}) : super(key: key);

 final String title;

 @override
 _MyHomePageState createState() => _MyHomePageState();
}

class _MyHomePageState extends State<MyHomePage> {
 int _counter = 0;

 void _incrementCounter() {
   setState(() {
     _counter++;
   });
 }

 @override
 Widget build(BuildContext context) {
   print("Building _MyHomePageState");
   //Widget#1
   return Scaffold(
     //Widget#2
     appBar: AppBar(
       title: Text(widget.title),
     ),
     //Widget#3
     body: Center(
```
```

```
    //Widget#5
    child: Column(
      mainAxisAlignment: MainAxisAlignment.center,
      children: <Widget>[
        //Widget#6
        Text(
          'You have pushed the button this many times:',
        ),
        //Widget#7
        Text(
          '$_counter',
          style: Theme.of(context).textTheme.headline4,
        ),
      ],
    ),
  ),
  //Widget#4
  floatingActionButton: FloatingActionButton(
    onPressed: _incrementCounter,
    tooltip: 'Increment',
    child: Icon(Icons.add),
  ),
);
}
}
```

As you can see from the code above, the stateful widget `MyHomePage` is assigned to the `home` property of MaterialApp. The `MyHomePage` has `_MyHomePageState` as its state widget responsible for managing the state for the widget. It's `build()` method builds the widgets. The seven widgets discussed above are used to build the *CounterApp*'s interface. Pressing on FAB's executes `_incrementCounter()` method to increase the counter by one. The `_counter` variable is keeping track of the count. Every time the method `_incrementCounter()` executes, it increases `_counter` by one. The counter is increased from `setState()` method. This method triggers the `build()` method. This behavior can be verified by adding a debug print statement `print("Building _MyHomePageState");`. Every time the widget `_MyHomePageState` is rebuilt, this print statement will be executed as well.

```
class _MyHomePageState extends State<MyHomePage> {
  int _counter = 0;
  void _incrementCounter() {
    setState(() {
      _counter++;
    });
  }
  @override
  Widget build(BuildContext context) {
```

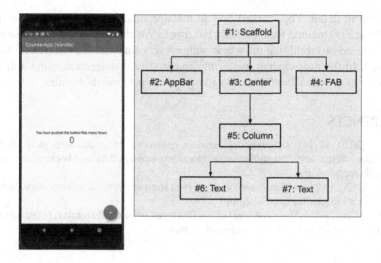

FIGURE 16.2 Visual for building vanilla implementation

```
print("Building _MyHomePageState");
return Scaffold(
  //appBar
  body: Center(
      ...
      Text('$_counter'),
      ...
      ),
    floatingActionButton: FloatingActionButton(
      onPressed: _incrementCounter,
      ...
    ),
  );
}
}
```

Figure 16.2 shows the widgets that get rebuilt every time the counter is updated, enclosed in a gray box.

In this simple app with less than ten widgets, managing state using the `set-State()` method would work without impacting performance. This approach is appropriate when managing a state within a single widget. However, in production apps, this approach for managing state may not work very well for large widgets. We'll learn managing state application-wide in upcoming chapters.

CONCLUSION

In this chapter, you learned about the types of states in a Flutter application's context. The default state-management solution for managing the ephemeral state is discussed in detail. You learned how to manage state in a single widget using the `setState()` method for `StatefulWidget`. We also discussed the performance issues caused by rebuilding the whole widget every time the `setState()` method is called. In the next chapter, we'll implement state management using a different approach known as *ValueNotifier* to address these performance issues.

REFERENCES

Google. (2020, 11 26). *Differentiate between ephemeral state and app state.* Retrieved from flutter.dev: https://flutter.dev/docs/development/data-and-backend/state-mgmt/ephemeral-vs-app

Google. (2020, 11 26). *setState method.* Retrieved from api.flutter.dev: https://api.flutter.dev/flutter/widgets/State/setState.html

Google. (2020, 11 26). *StatefulWidget class.* Retrieved from api.flutter.dev: https://api.flutter.dev/flutter/widgets/StatefulWidget-class.html

17 ValueNotifier

In the previous chapter (Chapter 16: Introduction to State Management), we discussed managing the widget's state using the stateful widget. You learned to manage state in StatefulWidget using the `setState()` method. We noticed that the interface is rebuilt every time the counter was updated inside the `setState()` method. In this chapter, we will look into one of the solutions to avoid rebuilding the entire widget tree of StatefulWidget. We will learn to use ValueNotifier (ValueNotifier<T> class) to only rebuild the Text widget displaying the counter value rather than rebuilding the entire widget tree. The widgets subscribed to the ValueNotifier get notified whenever its value is updated. Using ValueNotifier implementation for managing state helps to reduce the number of times a widget is rebuilt.

In this chapter, you will also learn to implement the state-management solution with the help of ValueNotifier, ValueListenable, ValueListenableBuilder widgets as described below:

1. ValueNotifier: This class belongs to Flutter's foundation library (foundation library). It extends ChangeNotifier (ChangeNotifier class). The ValueNotifier holds a single value at a time. You will learn more about ChangeNotifier in the next chapter (Chapter 18: Provider & ChangeNotifier).
2. ValueListenable (ValueListenable<T> class): This class belongs to Flutter's foundation library as well. It is the value emitted by ValueNotifier.
3. ValueListenableBuilder (ValueListenableBuilder<T> class): This widget listens to the value emitted by ValueNotifier. It gets rebuilt whenever the ValueNotifier emits a new value.

USING *ValueNotifier* APPROACH

Let's review the *CounterApp* widgets from the previous chapter. The app consists of Scaffold, AppBar, Center, FAB, Column, and Text widgets. Tapping the floating action button (FAB) is responsible for increasing the counter by one. In the vanilla (default) implementation, the `counter` variable is incremented inside the `setState()` method, which triggers rebuilding the whole interface at the Scaffold widget level.

In this chapter, we will learn to use ValueNotifier to implement incrementing the counter functionality. The implementation will make use of ValueNotifer, ValueListenable, and ValueListenableBuilder widgets.

In this implementation, the Text (Figure 17.1 – *Widget #8* in the app anatomy visual) widget to display the counter's value is wrapped inside the

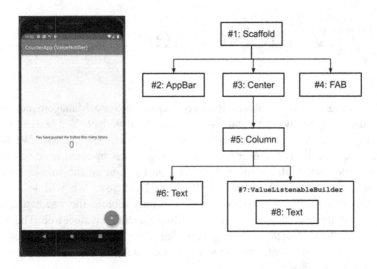

FIGURE 17.1   Anatomy of CounterApp for ValueNotifier implementation

ValueListenableBuilder (Figure 17.1 – *Widget #7* in the app anatomy visual) widget. There are a total of eight widgets in this app:

1. Scaffold: The root widget.
2. AppBar: Displays title of the app.
3. Center: Aligns its children in the center of the screen.
4. FloatingActionButton: FAB to increase the counter.
5. Column: Aligns its children vertically.
6. Text: Displays the information text.
7. ValueListenableBuilder: Wraps Text widget displaying the counter value.
8. Text: Displays the counter value.

Refer to Figure 17.1 for the visual structure of the app's widgets.

Let's dive into the code to understand the implementation details.

ValueNotifer

This class belongs to Flutter's foundation library. It extends ChangeNotifier. It holds a single value of any data type like `String`, `bool`, `int`, or a custom data type. Whenever its current value is replaced with a different value, it notifies its listeners.

The variable `_counter` is a ValueNotifier of type `int`. It holds the count that's displayed in the Text widget. The count is updated inside `_incrementCounter()` method. Whenever this method is executed, it notifies ValueNotifier's subscriber about this change.

```
class _MyHomePageState extends State<MyHomePage> {
```

ValueNotifier   259

```
final ValueNotifier<int> _counter = ValueNotifier<int>(0);
void _incrementCounter() {
  _counter.value += 1;
}
@override
Widget build(BuildContext context) {
  print("Building _MyHomePageState");
  return Scaffold(
    appBar: AppBar(
      title: Text(widget.title),
    ),
    body: Center(
      child: Column(
        mainAxisAlignment: MainAxisAlignment.center,
        children: <Widget>[
          Text(
            'You have pushed the button this many times:',
          ),
          //Builds when valueNotifier is changed/updated
          ValueListenableBuilder(
            builder: (BuildContext context, int value,
Widget child) {
              print("Building ONLY Text widget");
              return Text(
                '$value',
                style: Theme.of(context).textTheme.
headline4,
              );
            },
            valueListenable: _counter,
          ),
        ],
      ),
    ),
    floatingActionButton: FloatingActionButton(
      onPressed: _incrementCounter,
      tooltip: 'Increment',
      child: Icon(Icons.add),
    ),
  );
}
}
```

ValueListenable

This class (ValueListenableBuilder<T> class) belongs to Flutter's foundation library
as well. It holds the value emitted by ValueNotifier. The ValueListenable
exposes only one single current value at a given point in time. You depend on
ValueListenable's value to rebuild the widget tree. The `_counter` is the

value emitted by ValueNotifier when `_counter` is updated in `_incrementCounter()` method. The ValueListenable is notified about the updated `_counter`.

```
ValueListenableBuilder(
  builder: (BuildContext context, int value, Widget child) {
    print("Building ONLY Text widget");
    return Text(
      '$value',
      style: Theme.of(context).textTheme.headline4,
    );
  },
  valueListenable: _counter,
),
```

ValueListenableBuilder

This widget listens to the value emitted by ValueNotifier. It gets rebuilt whenever the ValueNotifier emits a new value. Its content is synced with ValueListenable. This widget registers itself with ValueListenable and calls its `builder` whenever an updated value is received.

The `ValueListenableBuilder` is called every time `_counter` is updated to a new value and rebuilds only *Widget #8* rather than building the StatefulWidget's entire widget tree.

```
//Mapping to Widget #7
ValueListenableBuilder(
  builder: (BuildContext context, int value, Widget child) {
    print("Building ONLY Text widget");
    //Widget #8
    return Text(
      '$value',
      style: Theme.of(context).textTheme.headline4,
    );
  },
  valueListenable: _counter,
),
```

COMPLETE CODE

Let's put together all the pieces in one place and run code in the IDE of your choice.

```
import 'package:flutter/material.dart';

void main() {
```

```
 runApp(CounterApp());
}


class CounterApp extends StatelessWidget {
// This widget is the root of your application.
@override
Widget build(BuildContext context) {
  return MaterialApp(
    debugShowCheckedModeBanner: false,
    theme: ThemeData(
      primarySwatch: Colors.blue,
      visualDensity: VisualDensity.adaptivePlatformDensity,
    ),
    home: MyHomePage(title: 'CounterApp (ValueNotifier)'),
  );
}
}

class MyHomePage extends StatefulWidget {
MyHomePage({Key key, this.title}) : super(key: key);
final String title;
@override
_MyHomePageState createState() => _MyHomePageState();
}
class _MyHomePageState extends State<MyHomePage> {
  final ValueNotifier<int> _counter = ValueNotifier<int>(0);
  void _incrementCounter() {
    _counter.value += 1;
  }
  @override
  Widget build(BuildContext context) {
    print("Building _MyHomePageState");
    //Widget#1
    return Scaffold(
      //Widget#2
      appBar: AppBar(
        title: Text(widget.title),
      ),
      //Widget#3
      body: Center(
        //Widget#5
        child: Column(
          mainAxisAlignment: MainAxisAlignment.center,
          children: <Widget>[
            //Widget#6
            Text(
              'You have pushed the button this many times:',
            ),
            //Widget#7. Builds when valueNotifier is changed/
updated
            ValueListenableBuilder(
```

```
        builder: (BuildContext context, int value, Widget
child) {
            print("Building ONLY Text widget");
            //Widget#8
            return Text(
              '$value',
              style: Theme.of(context).textTheme.headline4,
            );
          },
          valueListenable: _counter,
        ),
      ],
    ),
  ),
  //Widget#4
  floatingActionButton: FloatingActionButton(
    onPressed: _incrementCounter,
    tooltip: 'Increment',
    child: Icon(Icons.add),
  ),
 );
}
}
```

SOURCE CODE ONLINE

You can also review the full source code for this example (Chapter 17: ValueNotifier) at GitHub.

CONCLUSION

In this chapter, you learned to implement the state-management solution with `ValueNotifier`, `ValueListenable`, `ValueListenableBuilder` widgets. Using ValueNotifier implementation for managing state helps to reduce the number of times a widget is rebuilt. In the next chapter (Chapter 18: Provider & ChangeNotifier), you will be introduced to another approach for state management using the `provider` package.

REFERENCES

Google. (2020, 11 26). *ChangeNotifier class*. Retrieved from api.flutter.dev: https://api.flutter. dev/flutter/foundation/ChangeNotifier-class.html

Google. (2020, 11 26). *foundation library*. Retrieved from api.flutter.dev: https://api.flutter. dev/flutter/foundation/foundation-library.html

Google. (2020, 11 26). *ValueListenableBuilder<T> class*. Retrieved from api.flutter.dev: https://api.flutter.dev/flutter/widgets/ValueListenableBuilder-class.html

Google. (2020, 11 26). *ValueNotifier<T> class*. Retrieved from api.flutter.dev: https://api.flutter. dev/flutter/foundation/ValueNotifier-class.html

Tyagi, P. (2020, 11 26). *Chapter 17: ValueNotifier.* Retrieved from Pragmatic Flutter GitHub Repo: https://github.com/ptyagicodecamp/pragmatic_flutter/blob/master/lib/chapter17/valuenotifier_counterapp.dart

Tyagi, P. (2020, 11 26). *ValueListenable<T> class.* Retrieved from api.flutter.dev: https://api.flutter.dev/flutter/foundation/ValueListenable-class.html

Tyagi, P. (2021). Chapter 16: Introduction to State Management. In P. Tyagi, *Pragmatic Flutter: Building Cross-Platform Mobile Apps for Android, iOS, Web & Desktop.* CRC Press.

Tyagi, P. (2021). Chapter 18: Provider & ChangeNotifier. In P. Tyagi, *Pragmatic Flutter: Building Cross-Platform Mobile Apps for Android, iOS, Web & Desktop.* CRC Press.

18 Provider and ChangeNotifier

In this chapter, we will add one more approach to our exploration of state management. This approach will use a combination of `ChangeNotifier` (ChangeNotifier class) and `Provider` (provider) package to manage the counter state in *CounterApp*. This state-management solution help manages the state of widgets at the application level.

WHAT IS `Provider`

The `Provider` (provider) package is a wrapper around the `InheritedWidget` (InheritedWidget class) to make them easier to use and more reusable. The `InheritedWidget` is a base class for all widgets, and it efficiently propagates information down the widget tree. The provider(s) are placed above the widget(s) they're supposed to provide data.

In the *CounterApp*, the `CounterApp()` is added as a child to `ChangeNotifierProvider` widget. The `ChangeNotifierProvider` widget automatically manages `ChangeNotifier`. It creates `ChangeNotifier` (ChangeNotifier class) using `create`, and makes sure to dispose of it when `ChangeNotifierProvider` is removed from the widget tree. We will be implementing a custom `ChangeNotifier` named `CountNotifier()` later in this chapter.

```
ChangeNotifierProvider(
  create: (_) => CountNotifier(),
  child: CounterApp(),
),
```

ADDING DEPENDENCY

The provider package needs to be added to *pubspec.yaml* under the dependencies section. You may want to upgrade the provider package's version to the latest.

```
dependencies:
  flutter:
    sdk: flutter
  #Provider package
  provider: ^4.1.2
```

WHAT IS `ChangeNotifier`

The `ChangeNotifier` class belongs to Flutter's Foundation library (foundation library). It does what its name says. It notifies about the change to any widgets that are subscribed to it. It can either be extended or mixed using `mixins` (mixins) to provide a change notification API.

The *CounterApp* has `CountNotifier` that notifies about the update in the counter. It uses `notifyListeners()` method to do so. This method notifies all listeners/subscribers about the new update.

```
class CountNotifier with ChangeNotifier {
  int _counter = 0;
  int get counter => _counter;
  void incrementCounter() {
    print("Incrementing and notifying");
    _counter++;
    //Notifies about the change in counter
    notifyListeners();
  }
}
```

ANATOMY OF *CounterApp*

Now that we understand `ChangeNotifier` and `Provider` widgets' responsibilities, let's take a look at *CounterApp*'s widget tree in Figure 18.1.

Most of the widgets are arranged similar to examples discussed in previous chapters. There's a custom widget, `CountWidget`, responsible for displaying the updated counter value on the screen. There are a total of eight widgets in this app:

1. `Scaffold`: The root widget.
2. `AppBar`: Displays title of the app.
3. `Center`: Aligns its children in the center of the screen.
4. `FAB`: FloatingActionButton to increase the counter.
5. `Column`: Aligns its children vertically.
6. `Text`: Displays the information text.
7. `CountWidget`: A custom widget to display a counter value on the screen.
8. `Text`: It's the child to `CountWidget`. Displays the counter value.

Refer to Figure 18.1 for the anatomy of *CounterApp*'s implementation using Provider and ChangeNotifier.

Let's dive into the code to understand the implementation details. The *CounterApp*'s homepage is a `StatelessWidget`. This is because the state is managed using `ChangeNotifier` and `Provider`. The `ChangeNotifier` updates the value and notifies its subscribers. The provider package takes care of passing along this updated value to the appropriate widgets.

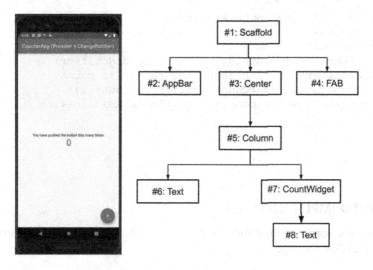

FIGURE 18.1 Anatomy of CounterApp for Provider and ChangeNotifier implementation

INCREASING COUNTER

The counter is increased by pressing the floating action button (FAB). Let's check out the implementation of `onPressed:` for this FAB. The `incrementCounter()` is accessed by calling **`read<CountNotifier>().incrementCounter()`** on the `context`. It executes the `incrementCounter()` method and updates the count without rebuilding the widget. The CountNotifier notifies its subscribers.

```
```
//Widget#4
floatingActionButton: FloatingActionButton(
 onPressed: () => context.read<CountNotifier>().
incrementCounter(),
 tooltip: 'Increment',
 child: Icon(Icons.add),
),
```
```

CUSTOM WIDGET: CountWidget

The CountWidget is the custom widget that displays the latest value of the `_ counter` on the screen. The Text widget watches for any broadcasts/notifications sent by the CountNotifier. As soon as it gets a new update, it gets rebuilt.

```
```
class CountWidget extends StatelessWidget {
 const CountWidget({Key key}) : super(key: key);
 @override
 Widget build(BuildContext context) {
```
```

```
    print("Building Count widget");
    //Widget#8
    return Text(
       //Rebuilds [CountWidget] when [CountNotifier]
ChangeNotifier notifies about the change in count
       '${context.watch<CountNotifier>().counter}',
       style: Theme.of(context).textTheme.headline4,
    );
  }
}
```

FINISHED IMPLEMENTATION

Here is the complete source code for the ChangeNotifier implementation with the Provider package.

```
import 'package:flutter/material.dart';
import 'package:provider/provider.dart';

//The CounterApp uses Provider package + ChangeNotifier
void main() {
  //Providers are above CounterApp.
  runApp(
    ChangeNotifierProvider(
      create: (_) => CountNotifier(),
      child: CounterApp(),
    ),
  );
}

//ChangeNotifier. Notifies about the change/update in counter
class CountNotifier with ChangeNotifier {
  int _counter = 0;
  int get counter => _counter;

  void incrementCounter() {
    print("Incrementing and notifying");
    _counter++;
    //Notifies about the change in counter
    notifyListeners();
  }
}

class CounterApp extends StatelessWidget {
  // This widget is the root of your application.
  @override
  Widget build(BuildContext context) {
    return MaterialApp(
```

```
      debugShowCheckedModeBanner: false,
      theme: ThemeData(
        primarySwatch: Colors.blue,
        visualDensity: VisualDensity.adaptivePlatformDensity,
      ),
      home: MyHomePage(title: 'CounterApp (Provider +
ChangeNotifier)'),
    );
  }
}

class MyHomePage extends StatelessWidget {
  final String title;
  MyHomePage({Key key, this.title}) : super(key: key);

  @override
  Widget build(BuildContext context) {
    print("Building MyHomePage widget");
    //Widget#1
    return Scaffold(
      //Widget#2
      appBar: AppBar(
        title: Text(this.title),
      ),
      //Widget#3
      body: Center(
        //Widget#5
        child: Column(
          mainAxisAlignment: MainAxisAlignment.center,
          children: <Widget>[
            //Widget#6
            Text(
              'You have pushed the button this many times:',
            ),

            //Widget#7
            //Extracting into separate widget helps it to
rebuild independently of [HomePage]
            const CountWidget(),
          ],
        ),
      ),
      //Widget#4
      floatingActionButton: FloatingActionButton(
        //Using read() instead of watch() helps NOT TO rebuild
the widget when there's change in count (or Counter
ChangeNotifier notifies)
        onPressed: () => context.read<CountNotifier>().
incrementCounter(),
        tooltip: 'Increment',
        child: Icon(Icons.add),
```

```
    ),
  );
 }
}


//Implementation of Widget#7
class CountWidget extends StatelessWidget {
 const CountWidget({Key key}) : super(key: key);
 @override
 Widget build(BuildContext context) {
   print("Building Count widget");
   //Widget#8
   return Text(
     //Rebuilds [CountWidget] when [CountNotifier]
ChangeNotifier notifies about the change in count
     '${context.watch<CountNotifier>().counter}',
     style: Theme.of(context).textTheme.headline4,
   );
 }
}
```

Source Code Online

The source code for this example (Chapter 18: Provider & ChangeNotifier) is available online on GitHub.

CONCLUSION

In this chapter, you learned to implement the state-management solution with the help of the `Provider` package and `ChangeNotifier` class. In this implementation, the homepage is a `StatelessWidget` since it's not managing the state directly but through `ChangeNotifier`. The `Provider` package is facilitating the state to pass along the children widgets.

REFERENCES

Flutter Team. (2020, 11 26). *provider*. Retrieved from pub.dev: https://pub.dev/packages/provider

Google. (2020, 11 26). *ChangeNotifier class*. Retrieved from api.flutter.dev: https://api.flutter.dev/flutter/foundation/ChangeNotifier-class.html

Google. (2020, 11 26). *foundation library*. Retrieved from api.flutter.dev: https://api.flutter.dev/flutter/foundation/foundation-library.html

Google. (2020, 11 26). *InheritedWidget class*. Retrieved from api.flutter.dev: https://api.flutter.dev/flutter/widgets/InheritedWidget-class.html

Tyagi, P. (2020, 11 26). *Chapter 18: Provider & ChangeNotifier*. Retrieved from Pragmatic Flutter GitHub Repo: https://github.com/ptyagicodecamp/pragmatic_flutter/blob/master/lib/chapter18/provider_changenotifier.dart

Tyagi, P. (2020, 11 26). *mixins*. Retrieved from dart.dev: https://dart.dev/guides/language/language-tour#adding-features-to-a-class-mixins

19 BLoC Design Pattern

In this chapter, we will explore BLoC Pattern to manage states in Flutter applications. The 'BLoC' stands for 'Business Logic Component'. The BLoC pattern was introduced by Google I/O 2018 (Build reactive mobile apps with Flutter (Google I/O '18)) that uses reactive programming to manage the data flow across the Flutter application(s). We will also learn to use the BLoC pattern for managing states in *CounterApp*.

WHAT IS BLoC PATTERN?

The BLoC is a design pattern to facilitate data flow to Flutter widgets and vice versa. It stands for *Business Logic Component*. The data sources can be API responses or data/events generated from widgets. The BLoC receives streams of the event(s) from data sources and/or widgets, perform business logic on events received and emits corresponding states. A BLoC architecture looks like as shown in Figure 19.1.

ANATOMY OF *CounterApp*

Now that we understand `ChangeNotifier` and `Provider` widgets' responsibilities from previous chapter (Chapter 18: Provider & ChangeNotifier), let's take a look at *CounterApp*'s widget tree in Figure 19.2.

Most of the widgets are arranged similar to example discussed in previous chapters. There are a total of seven widgets in this app:

1. `Scaffold`: The root widget.
2. `AppBar`: Displays title of the app.
3. `Center`: Aligns its children in the center of the screen.
4. `FAB`: FloatingActionButton to increase the counter.
5. `Column`: Aligns its children vertically.
6. `Text`: Displays the information text.
7. `Text`: Displays the counter value.

Widget #7 is the `Text` widget, which displays the updated counter. In the `setState()` approach, the whole widget tree was rebuilt. In the 'Provider & ChangeNotifier' implementation, the `Text` widget was notified about the counter change.

FIGURE 19.1   BLoC architecture overview

271

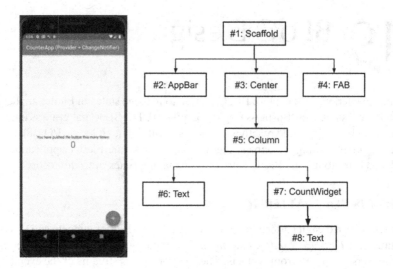

FIGURE 19.2   Anatomy of CounterApp for Provider and ChangeNotifier implementation

In this BLoC implementation, the BLoC's output state stream emits the state for a particular event, such as pressing the increment floating action button (FAB) to increase the counter. This increment event is fed into BLoC to generate a state for this event reflected in *Widget #7*, the `Text` widget.

Refer to Figure 19.3 for the anatomy of *CounterApp*'s implementation using the BLoC pattern.

In this chapter, we will explore three different implementations using the BLoC pattern.

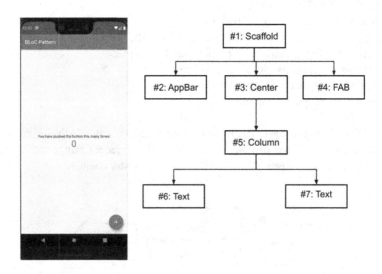

FIGURE 19.3   Anatomy of CounterApp's implementation using BLoC pattern

1. **Basic BLoC Pattern**: In this implementation, you will learn to implement a BLoC pattern using Dart's built-in data types like Stream (Stream<T> class), Sink (Sink<T> class), StreamController (StreamController<T> class) to manage the state of the counter in the default *CounterApp* application.
2. **Improvised BLoC Pattern**: In this implementation, you will improvise on the basic implementation by creating classes representing event(s) and state(s).
3. **BLoC Library**: In this implementation, you will learn to use the `flutter_bloc` (flutter_bloc) library to implement the BLoC pattern.

BASIC BLoC PATTERN IMPLEMENTATION

In this section, we will understand the fundamental and minimal implementation of the BLoC pattern. Let's look into the BLoC to understand how streams, sink, and stream controllers fit together to manage the state in the *CounterApp* Flutter application. First, let's understand the meaning of streams, sink, and stream controllers.

- Stream (Stream<T> class): A stream is a source of asynchronous data events. The Stream class provides a way to receive such a sequence of asynchronous data events. A stream is listened on to make it start generating events.
- Sink (Sink<T> class): The Sink class provides the destination for data. When an event is fired from the user interface, it can be put in a sink. A sink should be closed when no more data is being added to it.
- StreamController (StreamController<T> class): The StreamController class manages a stream. It creates a stream that can be listened to by others and adds events to it using Sink class.

Refer to Figure 19.4 to understand the BLoC internals.

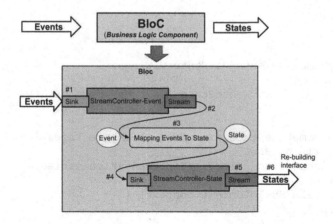

FIGURE 19.4 BLoC internal details

There are two streams in a BLoC. The first stream is to handle input events. In *CounterApp*, these input events are generated by user interaction with FAB. This *'event stream'* is managed by the *'StreamController-Event'*. The event stream sends these events to a logic box *'event to state'* mapper. This is where the events are handled as per the business logic and provides the appropriate state to be passed on to the widgets. The second stream handles these output states. The *'StreamController-State'* manages this output states stream. This *'state stream'* provides the output states to listening widgets to rebuild the specific widgets.

Let's understand the architecture shown in Figure 19.4 in the code implementation.

APP STRUCTURE

The *CounterApp* will use the BLoC pattern to manage the state of the counter. An increment counter event is added to the bloc's event stream. The bloc's state stream will provide the output state to update *Widget #7*. You will learn to implement the `CounterBloc` class next. The `CounterBloc` class is initiated in `_ MyHomePageState` as `final _ bloc = CounterBloc()`. It's important to close streams in `_ MyHomePageState`'s `dispose()` method to avoid memory leaks.

The overall *CounterApp* code structure is shown as below:

```
void main() {
 runApp(CounterApp());
}

class CounterApp extends StatelessWidget {
 @override
 Widget build(BuildContext context) {
   return MaterialApp(
     ...
     home: MyHomePage(title: 'BLoC Pattern'),
   );
 }
}

class MyHomePage extends StatefulWidget {
 @override
 _MyHomePageState createState() => _MyHomePageState();
}

class _MyHomePageState extends State<MyHomePage> {
 //Initializing BLoC
 final CounterBloc _bloc; //Implementation varies
 @override
 Widget build(BuildContext context) {
   return Scaffold(
     ...
     body: Center(
```

```
        child: Column(
          mainAxisAlignment: MainAxisAlignment.center,
          children: <Widget>[
            //Widget #6,
            //Building Widget #7 with updated state emitted by
BLoC's State stream
            //Widget #7,
          ],
        ),
      ),
      floatingActionButton: FloatingActionButton(
      //Sending Increment event to BLoC
      onPressed: () => _bloc.eventSink.add(CounterEvent.
increment),
        tooltip: 'Increment',
        child: Icon(Icons.add),
      ),
    );
  }

  @override
  void dispose() {
    super.dispose();
    //Closing bloc's stream controllers to avoid memory leaks
    _bloc.dispose();
  }
}
```
```

## CounterEvent

The `CounterEvent` enumeration is used to declare the event to increment the count. Whenever the FAB is pressed, a `CounterEvent.increment` event is pushed to the event stream's sink.

```
enum CounterEvent { increment }
```
```

CounterState

The *CounterApp* has only one state, which is the counter itself. In this case, the state can be represented with an `int _ counter`. The counter is '0' in the beginning, so it can be set to '0' as its initial state. It's declared inside `CounterBloc`.

```
Class CounterBloc {
    //Declaring state of the counter as int
    int _counter = 0;
}
```
```
```

CounterBloc

The CounterBloc class manages both the event and state streams with respective StreamController(s). It declares StreamController(s), Stream, and Sink for the event(s) and state(s). Let's look at these pieces one by one, which makes a CounterBloc class. Each step is marked with numbers in Figure 19.4.

```
Class CounterBloc {
    //#2. Listening to the event's stream in CounterBloc
constructor
    //Setting up event Stream, Sink & StreamController
    //Setting up state Stream, Sink & StreamController
    //#3. Mapping event to state (business logic)
    //Closing StreamControllers
}
```

EVENT StreamController

The _ eventController manages the input event stream and sink. This `StreamController` only supports the event type of `CounterEvent`, so generic is used to declare it.

```
final _eventController = StreamController<CounterEvent>();
```

EVENT Sink

User interface's interaction events are pushed into sink _ eventController's sink. The `eventSink` is returned using ` _ eventController.sink`.

```
Sink<CounterEvent> get eventSink => _eventController.sink;
```

EVENT Stream/CounterBloc CONSTRUCTOR

The input event Stream is being listened to, and the event(s) are retrieved. This step maps to '#2' in Figure 19.4. The event controller's stream is listening to the events and feeding into the event to the State mapper. It's listening to incoming UI events and mapping them into corresponding output States.

```
CounterBloc() {
 _eventController.stream.listen(
    (event) {
      _mapEventToState(event);
```

```
    },
  );
}
```

The same code above can also be written using lambda expression like below:

```
CounterBloc() {
    _eventController.stream.listen(_mapEventToState);
}
```

STATE `StreamController`

The State `StreamController` manages the stream of output states. The state is of int type, and it's declared as `StreamController<int>()`.

```
final _stateController = StreamController<int>();
```

STATE `Sink`

The output state generated by the event for state mapper will be added to the `_stateSink`. Details are yet to come in the **'Mapping Event to State'** section.

```
StreamSink<int> get _stateSink => _stateController.sink;
```

STATE `Stream`

The `_stateController` manages the stream of output states of int type. It provides the updated state(s) to the UI widgets. This step maps to the '#5' in Figure 19.4.

```
Stream<int> get counter => _stateController.stream;
```

MAPPING EVENT TO STATE

This `_mapEventToState(CounterEvent event)` method maps a given event to its corresponding state based on the business logic. When an event of increment type is passed, then the state of the _counter increases by one. This step maps to the '#3' in Figure 19.4.

This newly generated state is added into the `StateController`'s sink `_stateSink`. This step maps to the '#4' in Figure 19.4.

```
//#3. Mapping event to state
void _mapEventToState(CounterEvent event) {
 //Increment counter if event is matched
 if (event == CounterEvent.increment) _counter++;

 //#4: Adding output state to _stateSink
 _stateSink.add(_counter);
}
```

Closing `StreamController`

Don't forget to close the stream controllers for the event and state streams; otherwise, you'll get memory leaks in your app.

```
//close stream controllers to stop memory leak
void dispose() {
 _stateController.close();
 _eventController.close();
}
```

Initializing `CounterBloc`

The `CounterBloc` is initialized in the `_MyHomePageState` and `_bloc` holds the reference to it.

```
class _MyHomePageState extends State<MyHomePage> {
    final CounterBloc _bloc = CounterBloc();
}
```

Pushing UI Events

This step maps to the #1 in Figure 19.4. In this step, the user interaction with the app's FAB as `CounterEvent.increment` is added to the CounterBloc's `_eventSink`.

```
floatingActionButton: FloatingActionButton(
 //#1: Events added to eventController's sink
 onPressed: () => _bloc.eventSink.add(CounterEvent.increment),
 tooltip: 'Increment',
```

```
   child: Icon(Icons.add),
),
```
~ ~ ~

Rebuilding Widget/Consuming State

This is the last step of rebuilding the interface based on the generated output state for the given event. This step maps to the #6 in Figure 19.4. `StreamBuilder` builds the `Text` widget based on a new state received from `_bloc's StateController's` stream. The bloc's output stream is fed into `StreamBuilder` by assigning its `stream` property to `_bloc.counter`. The `snapshot` carries the output state, which is `int` in this case. The `snapshot.data` displays the updated counter value.

~ ~ ~
```
//#6: StreamBuilder(
 stream: _bloc.counter,
 initialData: 0,
 builder: (BuildContext context, AsyncSnapshot<int> snapshot)
{
    return Text(
      '${snapshot.data}',
      style: Theme.of(context).textTheme.headline4,
    );
},
),
```
~ ~ ~

Source Code Online

The source code for this example (Chapter 19: Basic BLoC) is available on GitHub.

IMPROVISED BLoC PATTERN IMPLEMENTATION

In this section, we will improvise the earlier implementation of declaring the events and states for the BLoC. In the last section, the event was declared using enumeration, and the state was declared as `int`. You'll learn to define event and state as a class implementation. Such implementation is useful when building complex apps where states/events are not trivial enough to represent as int or enumeration.

CounterEvent

The `CounterEvent` is declared as an `abstract` class. This is a good design practice when planning to support multiple different types of events handled by the app. In this case, you've only increment event, so define another subclass as `IncrementEvent` extending to `CounterEvent`. Whenever the FAB is pressed, an `IncrementEvent()` event is pushed to the event stream's sink.

```
` ` `
abstract class CounterEvent {}
class IncrementEvent extends CounterEvent {}
` ` `
```

CounterState

The counter's state is defined using a `CounterState` class. It has a `counter` variable of `int` type to hold the counter's value at a given time. The factory method `CounterState.initial()` provides the counter's state initialized to '0' by default.

```
` ` `
class CounterState {
  final int counter;
  const CounterState({this.counter});
  factory CounterState.initial() => CounterState(counter: 0);
}
` ` `
```

CounterBloc

The `CounterBloc` class remains the same as we discussed in the previous section. Let's discuss *only* the implementations that are different from the basic implementation above. There's no change in the event stream controller, sink, and the `CounterBloc()` constructor implementation.

STATE `StreamController`

The state `StreamController` manages the stream of output states. The state is of `CounterState` type, and it's declared as `StreamController<CounterState>()`.

```
` ` `
final _stateController = StreamController<CounterState>();
` ` `
```

STATE `Sink`

The output state generated by the event to state mapper will be added to the `_stateSink`. Note that the stream is of `CounterState` type as declared as `StreamSink<CounterState>`.

```
` ` `
StreamSink<CounterState> get _stateSink =>
_stateController.sink;
` ` `
```

STATE Stream

The `_stateController` manages the stream of output states of CounterState type. It provides the updated state(s) to the UI widgets. This step maps to the '#5' in Figure 19.4.

```
Stream<CounterState> get counter => _stateController.stream;
```

MAPPING EVENT TO STATE

This `_mapEventToState(CounterEvent event)` method maps a given event to its corresponding state based on the business logic. When an event `IncrementEvent()` is passed, then the current state `_currentState` gets updated to have counter incremented by one. This step maps to the '#3' in Figure 19.4.

This newly generated state is added into the StateController's sink `_stateSink`. This step maps to the '#4' in Figure 19.4.

```
//#3. Mapping event to state
void _mapEventToState(CounterEvent event) {
 //Increment counter if event is matched
 if (event is IncrementEvent) {
   _currentState = CounterState(counter: _currentState.counter
+ 1);
 }

  //#4: Adding current output state to _stateSink
  _stateSink.add(_currentState);
}
```

CLOSING StreamController

Closing stream controllers are similar to the previous implementation.

INITIALIZING CounterBloc

The `CounterBloc` initialization stays the same as before.

PUSHING UI EVENTS

This step maps to the #1 in Figure 19.4. In this step, the user interaction with the app's FAB as `IncrementEvent()` is added to the CounterBloc's `_eventSink`.

```
floatingActionButton: FloatingActionButton(
```

```
//#1: Events added to eventController's sink
onPressed: () => _bloc.eventSink.add(IncrementEvent()),
tooltip: 'Increment',
child: Icon(Icons.add),
),
~ ~ ~
```

REBUILDING WIDGET/CONSUMING STATE

This is the last step of rebuilding the interface based on the given event's generated output state. This step maps to the #6 in Figure 19.4. StreamBuilder builds the Text widget based on a new state received from _ bloc's StateController's stream. The bloc's output stream is fed into StreamBuilder by assigning its `stream` property to ` _ bloc.counter`. The `snapshot` carries the output state, which is `CounterState` in this case. The `snapshot.data.counter` displays the updated counter value.

```
~ ~ ~
//#6
StreamBuilder<CounterState>(
 stream: _bloc.counter,
 initialData: CounterState.initial(),
 builder:
     (BuildContext context, AsyncSnapshot<CounterState>
snapshot) {
   return Text(
     "${snapshot.data.counter}",
     style: Theme.of(context).textTheme.headline4,
   );
 },
),
~ ~ ~
```

SOURCE CODE ONLINE

The source code for this example (Chapter 19: Improvised BLoC) is available on GitHub.

IMPLEMENTING BLoC PATTERN USING LIBRARY

In this section, we will implement the BLoC pattern using a Flutter package/library `flutter _ bloc` (flutter_bloc) available at pub.dev (Flutter & Dart Packages). At the time of this writing, the version of `flutter _ bloc` library is `6.0.1`. It's added in *pubspec.yaml* as a dependency as shown below:

```
~ ~ ~
#flutter_bloc package
flutter_bloc: ^6.0.1
~ ~ ~
```

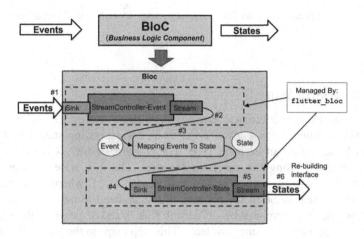

FIGURE 19.5   BLoC using 'flutter_bloc' library

The reason behind using this library is to reduce the boiler-plate code. The library provides the abstraction for managing events and state streams. Refer to <bloclib_ arch.jpg> to see the parts of the BLoC managed by the *'flutter_bloc'* (flutter_bloc) library (Figure 19.5).

As you see in the diagram above, the library provides stream controllers, sinks, and streams management. The developers must manage to push events to the bloc, mapping/business logic, and handle the states emitted by the bloc.

The `CounterEvent`, `IncrementEvent`, `CounterState` classes remain the same as of previous section. Let's check out the differentiated implementation below.

CounterBloc

The library manages `CounterBloc` implementation. The `CounterBloc` extends the `Bloc` class provided by the library. The event type-`CounterEvent` and state type-`CounterState` are passed as the generics. Step #2 shown in the diagram above is taken care of by the class definition itself. The *bloc* library provides automatic stream management for us.

The initial state of the `CounterState` is passed in the `CounterBloc` constructor.

```
```
//#2: Stream management
class CounterBloc extends Bloc<CounterEvent, CounterState> {
    //Initializing initial CounterState
    CounterBloc(CounterState initialState) :
super(initialState);
    Stream<CounterState> mapEventToState(CounterEvent event)
async* {
```
```

```
        //Implementation in Mapping Event to State section
    }.
}
~ ~ ~
```

MAPPING EVENT TO STATE

The `Stream<CounterState> mapEventToState(CounterEvent event) async*{}` method maps a given event to its corresponding state based on the business logic. When an event of type `IncrementEvent` is passed, the new state is generated using the generator function. The new state is calculated as `CounterState(counter: state.counter + 1)` where `state.counter` is the current value of the counter. It gets updated to have a counter incremented by one. The combination of `async*` & `yield` provides the stream of CounterState as it becomes available. This step maps to the '#3' in the diagram Figure 19.5. Internally, this step maps to the '#4' & '#5' in Figure 19.5.

```
~ ~ ~

//#3: Mapping events to their corresponding state based on the
business logic
@override
Stream<CounterState> mapEventToState(CounterEvent event)
async* {
  if (event is IncrementEvent) {
    //#4 + #5: Stream<State> provides the stream of state.
Heavy lifting done by BLoC library
    yield CounterState(counter: state.counter + 1);
  }
}
~ ~ ~
```

CLOSING `StreamController`

Closing stream controllers is the same as in the previous implementations.

INITIALIZING `CounterBloc`

The `CounterBloc` initialization is similar to previous examples with a twist. The `CounterBloc` constructor takes the initial state for the `CounterState`.

```
~ ~ ~

final CounterBloc _bloc = CounterBloc(CounterState.initial());
~ ~ ~
```

PUSHING UI EVENTS

This step maps to the #1 in Figure 19.5. In this step, the user interaction with the app's FAB as `IncrementEvent()` is added to the `CounterBloc` and managed by the library thereafter.

```
```
floatingActionButton: FloatingActionButton(
 //#1: Events added to eventController's sink
 onPressed: () => _bloc.add(IncrementEvent()),
 tooltip: 'Increment',
 child: Icon(Icons.add),
),
```
```

REBUILDING WIDGET/CONSUMING STATE

This is the last step of rebuilding the interface based on the generated output state for the given event. This step maps to the #6 in Figure 19.5. The `BlocBuilder` from the library is used to listen to the stream of output `CounterState`. Its `cubit` property is assigned to an instance of `CounterBloc` (_ bloc). The widget Text is built inside `builder`, which provides the current `CounterState` to display the content.

```
```
//#6
BlocBuilder<CounterBloc, CounterState>(
 cubit: _bloc,
 builder: (context, state) {
   return Text(
     '${state.counter}',
     style: Theme.of(context).textTheme.headline4,
   );
 },
)
```
```

SOURCE CODE ONLINE

The source code for (Chapter 19: Using BLoC Library) is available on GitHub.

CONCLUSION

In this chapter, you learned to implement the state-management solution with the BLoC pattern using in-built dart features like Stream, Sink, and StreamController. You also learned to improvise the implementations by converting events and state into their classes. Lastly, the *bloc* library is explored to reduce the boiler-plate code due to manual implementation.

REFERENCES

Dart Team. (2020, 11 26). *Sink<T> class*. Retrieved from api.dart.dev: https://api.dart.dev/stable/2.8.4/dart-core/Sink-class.html

Dart Team. (2020, 11 26). *Stream<T> class*. Retrieved from api.dart.dev: https://api.dart.dev/stable/2.8.4/dart-async/Stream-class.html

Dart Team. (2020, 11 26). *StreamController<T> class*. Retrieved from api.dart.dev: https://
api.dart.dev/stable/2.8.4/dart-async/StreamController-class.html

Flutter & Dart Team. (2020, 11 26). *Flutter & Dart Packages*. Retrieved from pub.dev:
https://pub.dev/

Flutter Team. (2020, 11 26). *flutter_bloc*. Retrieved from pub.dev: https://pub.dev/packages/
flutter_bloc

Tyagi, P. (2020, 11 26). *Chapter 19: Basic BLoC*. Retrieved from Pragmatic Flutter GitHub
Repo: https://github.com/ptyagicodecamp/pragmatic_flutter/blob/master/lib/chapter19/
bloc_pattern1.dart

Tyagi, P. (2020, 11 26). *Chapter 19: Improvised BLoC*. Retrieved from Pragmatic Flutter
GitHub Repo: https://github.com/ptyagicodecamp/pragmatic_flutter/blob/master/lib/
chapter19/bloc_pattern2.dart

Tyagi, P. (2020, 11 26). *Chapter 19: Using BLoC Library*. Retrieved from Pragmatic Flutter
GitHub Repo: https://github.com/ptyagicodecamp/pragmatic_flutter/blob/master/lib/
chapter19/flutter_bloc.dart

Tyagi, P. (2021). Chapter 18: Provider & ChangeNotifier. In P. Tyagi, *Pragmatic Flutter:
Building Cross-Platform Mobile Apps for Android, iOS, Web & Desktop*. CRC Press.

YouTube. (2020, 11 26). *Build reactive mobile apps with Flutter (Google I/O '18)*. Retrieved
from YouTube: https://www.youtube.com/watch?v=RS36gBEp8OI

20 Unit Testing

In this chapter, you will learn the basics of unit testing (Unit Testing) in a Flutter application. Unit tests are useful for testing a particular functionality of a method/ function or class. The modified version of default counter application is used to learn writing unit tests for a Flutter application. In the modified default *CounterApp*, we will add two buttons. We will use a Flutter plugin font _ awesome _ flutter (font_awesome_flutter) for the increment and decrement button icons. This plugin comes with a rich set of Flutter icons. We will add one green button with an increment icon for increasing the counter by one. Another red button with a decrement icon is used for reducing the counter value by one.

Once this modified interface for CounterApp is ready, you will learn how to write unit tests to test the logical functions for the given Dart class 'Counter'. The default *CounterApp* is improvised, as shown in Figure 20.1 to demonstrate testing in Flutter applications.

PACKAGE DEPENDENCIES

The test (test package) package is enough to write unit tests for pure Dart code. However, the flutter _ test (flutter_test) package is the right choice when testing Flutter applications. It also comes with tools to test Flutter widgets, which we will be exploring in the next chapter (Chapter 21: Widget Testing). In this chapter, you have the option to add either the test or flutter _ test package to *pubspec.yaml*. The test package is the minimum required dependency for unit tests. The test is added under the `dev _ dependencies` section of the *pubspec.yaml* file.

```
` ` `
dev_dependencies:
  test: ^1.15.3
` ` `
```

The font _ awesome _ flutter is added under the `dependencies` section of the *pubspec.yaml* file.

```
` ` `
dependencies:
  font_awesome_flutter: ^8.8.1
` ` `
```

THE *CounterApp*

In this section, we'll build each part of the app one by one starting from Counter pure Dart class.

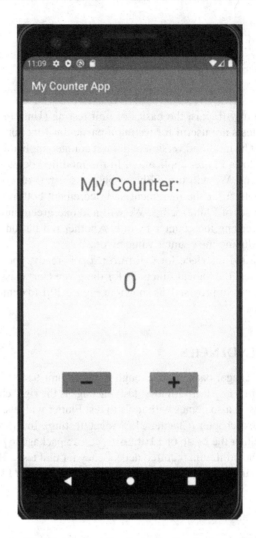

FIGURE 20.1   Improvised CounterApp for demonstrating testing in Flutter

Counter CLASS

The Counter class contains the current value of the number in the `number`
variable. The `increment()` method increases the `number` by one, while the
`decrement()` method decreases the `number` by one.

```
class Counter {
 int number = 0;
 void increment() => number++;
 void decrement() => number--;
}
```

CounterApp STATELESSWIDGET

The `CounterApp` extends `StatelessWidget`. It's the root level widget that works as a container for its descendent widgets. The `MaterialApp` is returned in its `build()` method. The default theme is applied to the `MaterialApp` as well. The homepage for the app is assigned using the `home` property of the `MaterialApp`.

```
class CounterApp extends StatelessWidget {
  @override
  Widget build(BuildContext context) {
    return MaterialApp(
      debugShowCheckedModeBanner: false,
      theme: ThemeData(
        primarySwatch: Colors.blue,
        visualDensity: VisualDensity.adaptivePlatformDensity,
      ),
      home: MyHomePage(title: 'My Counter App'),
    );
  }
}
```

MyHomePage STATEFULWIDGET

The `StatefulWidget` keeps track of the state of the variable(s) in the widget. The `MyHomePage` is a `StatefulWidget`. It uses the `_MyHomePageState` class to hold the state(s) of the variables. It keeps a reference to the `Counter` class we created earlier to access the counter's number.

```
class MyHomePage extends StatefulWidget {
  MyHomePage({Key key, this.title}) : super(key: key);
  final String title;
  @override
  _MyHomePageState createState() => _MyHomePageState();
}
class _MyHomePageState extends State<MyHomePage> {
  Counter _counter = Counter();
}
```

METHOD TO INCREMENT COUNTER NUMBER

The `_MyHomePageState` class has a private method to increment the counter by one. It calls `_counter.increment()` method from `setState()` method to trigger rebuild the widgets to render the current counter number on the screen.

```
` ` `
void _incrementCounter() {
 setState(() {
   _counter.increment();
 });
}
` ` `
```

METHOD TO DECREMENT COUNTER NUMBER

The `_decrementCounter()` method is similar to `_increment-Counter()` method but decrements the counter number by one.

```
` ` `
void _decrementCounter() {
 setState(() {
   _counter.decrement();
 });
}
` ` `
```

MyTextWidget STATELESSWIDGET

Let's create a reusable custom Text widget to display all the Text widgets throughout the app. This custom widget takes a string to be displayed, TextStyle to be applied to this widget, and Key to preserving the state of the widget. We also need to specify the directionality of the Text widget using TextDirection (TextDirection enum) because this widget will be tested independently as a unit. The `textDirection` property of the Text widget is assigned to `TextDirection.ltr`. However, you don't need to specify textDirection in your Text widget when it's a descendent of the Scaffold widget because textDirection is implicitly provided in that case.

```
` ` `
class MyTextWidget extends StatelessWidget {
 final String textString;
 final TextStyle style;
 final Key myKey;
 const MyTextWidget({this.myKey, this.textString, this.
style});
 @override
 Widget build(BuildContext context) {
   return Text(
     textString,
     style: style,
     textDirection: TextDirection.ltr,
     key: myKey,
   );
 }
}
` ` `
```

Displaying Label

The MyTextWidget custom widget is used to display the label 'My Counter:' as below:

```
MyTextWidget (
 myKey: Key('L'),
 textString: 'My Counter:',
 style: Theme.of(context).textTheme.headline4,
),
```

This widget is added as the first child to the centered Column widget.

Displaying Number

The MyTextWidget is also used to display the current counter number. The widget's key is `Key('L')`. The current counter number is assigned to the `textString` property of MyTextWidget. It's accessed as `${ _ counter.number}`.

Displaying Decrement Button

The `ElevatedButton` (ElevatedButton class) widget is used to decrement the counter's number by one. Its `key` property is assigned to `Key('D')`. It's also assigned a red color using `Colors.red`. The minus icon is applied using the `font _ awesome _ flutter` Flutter plugin. The icon `FaIcon(FontAwesomeIcons. minus)` is assigned as the child to the `ElevatedButton` widget. The `onPressed()` method calls the `_ decrementCounter()` to decrement the counter number by one.

```
ElevatedButton (
 key: Key('D'),
 style: ElevatedButton.styleFrom(primary: Colors.red),
 onPressed: () => _decrementCounter(),
 child: FaIcon(FontAwesomeIcons.minus),
),
```

Displaying Increment Button

The increment button is assigned to the key `Key('I')`, and green color. The ElevatedButton widget's child is `FaIcon(FontAwesomeIcons.plus)`. The `onPressed()` method increases the counter number by one.

```
ElevatedButton (
 key: Key('I'),
```

```
  style: ElevatedButton.styleFrom(primary: Colors.green),
  onPressed: () => _incrementCounter(),
  child: FaIcon(FontAwesomeIcons.plus),
)
```

TEST FILE

The next step is to create a test file to test the counter number's increment and decrement functionality. Usually, this test file is created under the *'test'* folder of the
root of the Flutter project. The file name ends in the *'_test.dart'* suffix. Since our
CounterApp is located in *'lib/chapter20/main_20.dart'*, the test file will be *'test/
main_20_test.dart'*. The tests are written in a `test` block like below:

```
test('Test description', () {
 //test code
});
```

Multiple tests can be combined in one group like below:

```
group('Test group, () {
 test('Test#1 Description', () {});
 test('Test#2 Description', () {});
});
```

We'll be testing the `Counter` class for four scenarios: default number at the beginning, increasing the number by one, decreasing the number by one, and finally incrementing and decrementing number in order.

TESTING DEFAULT NUMBER

In this test case, the default counter number is tested. It's expected to have a default number as '0'. The number is accessed from the `Counter` class as `counter.number`.

```
test('Default Number is 0', () {
 final counter = Counter();
 expect(counter.number, 0);
});
```

INCREMENTING NUMBER

Increasing the number by one will increase the counter by one. In this case, it's
expected `counter.number` to be equal to '1'.

```
~ ~ ~
test('Increment number by 1', () {
 final counter = Counter();
 counter.increment();
 expect(counter.number, 1);
});
~ ~ ~
```

Decrementing Number

Decreasing a number by one will decrement the default counter '0' by one, which is '–1'. It's expected `counter.number` to be equal to '–1' in this case.

```
~ ~ ~
test('Decrement number by 1', () {
 final counter = Counter();
 counter.decrement();
 expect(counter.number, -1);
});
~ ~ ~
```

Increment & Decrement Together

In this test case, the number is first incremented by one and then decremented by one resulting in the final number as '0'.

```
~ ~ ~
test('Increment & Decrement number by 1', () {
 final counter = Counter();
 counter.increment();
 expect(counter.number, 1);
 counter.decrement();
 expect(counter.number, 0);
});
~ ~ ~
```

Complete Source Code

```
~ ~ ~
import 'package:pragmatic_flutter/chapter24/main_24.dart';
import 'package:test/test.dart';
//Unit Testing: CounterApp
void main() {
 group('CounterApp-Unit Tests', () {
   test('Default Number is 0', () {
     final counter = Counter();
     expect(counter.number, 0);
   });

   test('Increment number by 1', () {
     final counter = Counter();
```

```
    counter.increment();
    expect(counter.number, 1);
  });

  test('Decrement number by 1', () {
    final counter = Counter();
    counter.decrement();
    expect(counter.number, -1);
  });

  test('Increment & Decrement number by 1', () {
    final counter = Counter();
    counter.increment();
    expect(counter.number, 1);

    counter.decrement();
    expect(counter.number, 0);
  });
 });
}
```

SOURCE CODE ONLINE

The full source code of this example (Chapter20: Unit Testing) is available at GitHub.

RUNNING UNIT TESTS

There are two ways to run unit tests. One way is to right-click on the test file and the second way is to choose the option to run tests from the pop-up menu. Tests can also be run from the command line using the following command:

```
flutter test test/<testfile-suffixed-with_test>.dart
```

Let's run the test file for the current example with the command below:

```
flutter test test/main_20_test.dart
```

OUTPUT

The output shows the time taken to execute all the tests, followed by the number of tests completed and the message on whether tests were passed.

```
00:14 +4: All tests passed!
```

CONCLUSION

In this chapter, you learned how to test the logical functionalities for the given Dart `Counter` class. In the next chapter, you will learn to test the widgets that compose the user interface.

REFERENCES

dart.dev. (2020, 11 27). *test package*. Retrieved from pub.dev: https://pub.dev/packages/test

Flutter Community. (2020, 11 27). *font_awesome_flutter*. Retrieved from pub.dev: https://pub.dev/packages/font_awesome_flutter

Flutter Team. (2020, 11 27). *flutter_test*. Retrieved from flutter.dev: https://flutter.dev/docs/cookbook/testing/widget/introduction#1-add-the-flutter_test-dependency

Google. (2020, 11 27). *TextDirection enum*. Retrieved from flutter.dev: https://api.flutter.dev/flutter/dart-ui/TextDirection-class.html

Google. (2020, 11 30). *ElevatedButton class*. Retrieved from api.flutter.dev: https://api.flutter.dev/flutter/material/ElevatedButton-class.html

Google. (2020, 11 30). *Unit Testing*. Retrieved from flutter.dev: https://flutter.dev/docs/cookbook/testing/unit

Tyagi, P. (2020, 11 30). *Chapter20: Unit Testing*. Retrieved from Pragmatic Flutter GitHub Repo: https://github.com/ptyagicodecamp/pragmatic_flutter/blob/master/test/main_20_test.dart

Tyagi, P. (2021). Chapter 21: Widget Testing. In P. Tyagi, *Pragmatic Flutter: Building Cross-Platform Mobile Apps for Android, iOS, Web & Desktop*. CRC Press.

21 | Widget Testing

In this chapter, you will learn the basics of testing widgets (Widget Testing) in a Flutter application. The widget test lets build and test widgets interactively. We will use the improvised default *CounterApp* example from the previous chapter (Chapter 21: Unit Testing). Earlier, we used the *CounterApp* to learn writing unit tests for Flutter applications. In this chapter, you will use the same *CounterApp* to learn writing tests for this app;s widgets. We will test the custom Text widget `MyTextWidget` used for displaying label text and counter number. Additionally, we will write a widget test case for the `ElevatedButton` widget to increment and decrement the number on the screen.

PACKAGE DEPENDENCY

In the last chapter, the test (test) package was enough to write unit tests for pure Dart code. For testing Flutter widgets, you need tools shipped with the `flutter _ test` (flutter_test) package. The `flutter _ test` dependency is automatically added to *pubspec.yaml* under the `dev _ dependencies` section when creating the Flutter project.

```
dev_dependencies:
  flutter_test:
    sdk: flutter
```

The `flutter _ test` package comes with tools useful to widget testing. We will use the following tools to write widget tests:

1. WidgetTester (WidgetTester class): The WidgetTester let's build interactive test widgets. It interacts with widgets and test environments. These widgets can be tapped programmatically to mimic the real-world user interface interactions. We will be using the following methods of WidgetTester class:
   a. pumpWidget(Widget widget) (pumpWidget method): This method builds the given widget `widget` in a test environment.
   b. pump() (pump method): This method rebuilds the already built widgets in the test environment. This is a useful method when testing StatefulWidgets, which needs to be rebuilt when their state is modified.
   c. tap(Finder finder) (tap method): This method simulates the tapping on a widget in the test environment on the found widget using Finder (Finder class) class.

2. Function testWidgets (testWidgets function): The `testWidgets()` functions are similar to `test()` functions used in unit tests. This is where you write a test code for verifying a widget's behavior. It creates a WidgetTester for the given test case.

3. Finder (Finder class): It helps search for widgets in the test app. The Finder class searches a widget tree to find a match for the pattern(s) and returns the matched node(s).

4. Matcher (Matcher class): This helps verify whether the Finder class was able to locate a given widget in the test environment.

TESTING WIDGETS

We will use all tools and functions mentioned above to write widget tests for *CounterApp*. The MyTextWidget custom Text widget is used to render the display label as well as to render the current counter number. The `textWidgets()` function is used to write test code. It provides a `WidgetTester` class to create widgets in the test environment. The `testWidget()` function is executed asynchronously and doesn't block another test case's execution. The widgets to be tested are created using the `pumpWidget()` method asynchronously.

CUSTOM Text WIDGET (LABEL)

In this test, we will be testing the widget rendering the 'My Counter:' text label on the screen. The `testWidgets()` function is used to write the test code for testing custom widget MyTextWidget with `textString` property. This widget is created using WidgetTester provided by the `testWidgets` function. The WidgetTester class uses the `pumpWidget()` method to create MyTextWidget widget with `textString` property assigned to `My Counter:` string. Once the widget to be tested is created, the Finder class method `find()` is used to find the occurrence of a `Text` widget rendering 'My Counter:' string. The matched widget in the widget tree is stored in the `myCounterText` variable. The `Matcher` constant `findsOneWidget` matches *one* widget that meets the criteria. The `expect()` function asserts if `myCounterText` matches the `findsOneWidget` matcher.

```
```
testWidgets('Testing Label Text Widget (My Counter:)',
    (WidgetTester widgetTester) async {
  //Building Text widget using WidgetTester
  await widgetTester.pumpWidget(MyTextWidget(textString: 'My
Counter:'));

  //Creating Finder to widget created
  final myCounterText = find.text('My Counter:');

  //Verifying widget using Matcher constant
  expect(myCounterText, findsOneWidget);
});
```
```

Custom **Text** Widget (number)

The counter number is displayed as a string using MyTextWidget on the screen. The MyTextWidget is created in a test environment with a default `textString` property initialized to '0'. The `widget.pumpWidget()` method creates the MyTextWidget widget. The `find()` method finds the Text widget rendering '0' text. Since there's only one Text widget rendering '0', the `Matcher` constant `findsOneWidget` is used to assert using the `expect()` method as shown in the code below:

```
testWidgets('Testing Counter Text Widget for number 0',
  (WidgetTester widgetTester) async {

 //Creating Text widget
 await widgetTester.pumpWidget(MyTextWidget(textString: '0'));
 //Creating Finder to widget created
 final myCounterText = find.text('0');

 //Verifying widget using Matcher constant
 expect(myCounterText, findsOneWidget);
});
```

Finding Widget Using Key

The other way to find a widget in the widget tree is by using the widget key. A Key (Key class) is an identifier to identify widgets in a widget tree. We will create the test widget using Key. We will run the test for the counter number the same as in the last test case, but using the Key this time. The `widgetKey` is created using `Key('T')`, and assigned to the custom widget MyTextWidget using the `mKey` property. Refer to the last chapter on Unit testing to review the *CounterApp* in detail. The MyTextWidget is found using the same `find()` method but with a key `widgetKey` created earlier: `find.byKey(widgetKey)`, and store its reference in `myCounterText` variable. Finally, we assert the `myCounterText` with the `findsOneWidget` matcher.

```
testWidgets('Finding widget by Key', (WidgetTester
widgetTester) async {
 final widgetKey = Key('T');
 //Creating Text widget
 await widgetTester
.     pumpWidget(MyTextWidget(myKey: widgetKey, textString:
'0'));

 //Creating Finder to widget created
 final myCounterText = find.byKey(widgetKey);

 //Verifying widget using Matcher constant
```

```
expect(myCounterText, findsOneWidget);
});
```

INCREMENT **ElevatedButton** WIDGET

The green button with the plus icon increases the counter number by one each time it's pressed. First, we create the app `CounterApp` in the test environment like below using `MediaQuery` (MediaQuery class) and adding `CounterApp()` as its descendent. The `MediaQuery` helps establish a subtree in which media queries resolve to the given data. The `MediaQueryData` (MediaQuery class) is needed to get the media information, like the size of the device the app is rendered on. The test app is created using `pumpWidget()` method by passing on the `testWidget` variable.

```
Widget testWidget = MediaQuery(
 data: MediaQueryData(),
 child: CounterApp(),
);
await widgetTester.pumpWidget(testWidget);
```

Next, we want to find the increment button and press it to increment the counter number. The increment `ElevatedButton` widget is created using key `I` as below:

```
ElevatedButton(
 key: Key('I'),
 style: ElevatedButton.styleFrom(primary: Colors.green),
 onPressed: () => _incrementCounter(),
 child: FaIcon(FontAwesomeIcons.plus),
)
```

The increment button is found using `find.byKey(Key('I'))`. This widget is tapped on using `await widgetTester.tap(.)`. The `pump()` method needs to be called to propagate the action to the widget tree.

```
await widgetTester.tap(
 find.byKey(Key('I')),
);
```

The number counter will be incremented by one. This is tested by finding the `Text` widget displaying the number as '1' and matching it with the `findsOne-Widget` matcher. The full incrementing counter button widget testing code is as below:

```
```
testWidgets('Increment Number', (WidgetTester widgetTester)
async {
 Widget testWidget = MediaQuery(
   data: MediaQueryData(),
   child: CounterApp(),
 );
 await widgetTester.pumpWidget(testWidget);

 //Tap increment Button (Green)
 await widgetTester.tap(
   find.byKey(Key('I')),
 );

 await widgetTester.pump();

 //Creating Finder to widget created
 final myCounterText = find.text('1');

 //Verifying widget using Matcher constant
 expect(myCounterText, findsOneWidget);
});
```
```

DECREMENT ElevatedButton WIDGET

The red button with the minus icon decrements the counter number by one every time it's pressed. In this test case, the tapping button will decrease the number by one. The default starting number is '0'. If it's decremented by one, then the number counter will be '–1'. The test case is below:

```
```
testWidgets('Decrement Number', (WidgetTester widgetTester)
async {
 Widget testWidget =
     MediaQuery(data: MediaQueryData(), child: CounterApp());
 await widgetTester.pumpWidget(testWidget);

 //Tap increment Button (Red)
 await widgetTester.tap(find.byKey(Key('D')));

 await widgetTester.pump();

 //Creating Finder to widget created
 final myCounterText = find.text('-1');

 //Verifying widget using Matcher constant
 expect(myCounterText, findsOneWidget);
});
```
```

INCREMENTING AND DECREMENTING

In this test case, let increment and decrement counter number one at a time. The starting number rendered as '0'. Incrementing the counter by one will render the number as '1'. Tapping on a button to decrease the number by one will decrement it by one. The result will be '0' again. The test case is below:

```
```
testWidgets('Incrementing & Decrementing Number',
    (WidgetTester widgetTester) async {
  Widget testWidget =
      MediaQuery(data: MediaQueryData(), child:
CounterApp());
    await widgetTester.pumpWidget(testWidget);

    //Tap increment Button (Green)
    await widgetTester.tap(find.byKey(Key('I')));

    await widgetTester.pump();

    //Tap increment Button (Red)
    await widgetTester.tap(find.byKey(Key('D')));

    await widgetTester.pump();

    //Creating Finder to widget created
    final myCounterText = find.text('0');

    //Verifying widget using Matcher constant
    expect(myCounterText, findsOneWidget);
  });
});
```
```

COMPLETE SOURCE CODE

The full source code of widget tests for *CounterApp* is below:

```
```
import 'package:flutter/material.dart';
import 'package:flutter_test/flutter_test.dart';
import 'package:pragmatic_flutter/chapter25/main_25.dart';

//Widget Testing: CounterApp
void main() {
 group("CounterApp-Widget Testing", () {
    testWidgets('Testing Label Text Widget (My Counter:)',
        (WidgetTester widgetTester) async {
      //Building Text widget using WidgetTester
      await widgetTester.pumpWidget(MyTextWidget(textString:
'My Counter:'));
```
```

```
        //Creating Finder to widget created
        final myCounterText = find.text('My Counter:');

        //Verifying widget using Matcher constant
        expect(myCounterText, findsOneWidget);
    });
    testWidgets('Testing Counter Text Widget for number 0',
        (WidgetTester widgetTester) async {
        //Creating Text widget
        await widgetTester.pumpWidget(MyTextWidget(textString:
'0'));

        //Creating Finder to widget created
        final myCounterText = find.text('0');

        //Verifying widget using Matcher constant
        expect(myCounterText, findsOneWidget);
    });

    testWidgets('Finding widget by Key', (WidgetTester
widgetTester) async {
        final widgetKey = Key('N');
        //Creating Text widget
        await widgetTester
            .pumpWidget(MyTextWidget(myKey: widgetKey,
textString: '0'));

        //Creating Finder to widget created
        final myCounterText = find.byKey(widgetKey);

        //Verifying widget using Matcher constant
        expect(myCounterText, findsOneWidget);
    });

·   testWidgets('Increment Number', (WidgetTester widgetTester)
async {
        Widget testWidget = MediaQuery(
          data: MediaQueryData(),
          child: CounterApp(),
        );
        await widgetTester.pumpWidget(testWidget);

        //Tap increment Button (Green)
        await widgetTester.tap(
          find.byKey(Key('I')),
        );

        await widgetTester.pump();

        //Creating Finder to widget created
```

```
    final myCounterText = find.text('1');

    //Verifying widget using Matcher constant
    expect(myCounterText, findsOneWidget);
  });

  testWidgets('Decrement Number', (WidgetTester widgetTester)
async {
    Widget testWidget =
        MediaQuery(data: MediaQueryData(), child:
CounterApp());
    await widgetTester.pumpWidget(testWidget);

    //Tap decrement Button (Red)
    await widgetTester.tap(find.byKey(Key('D')));

    await widgetTester.pump();

    //Creating Finder to widget created
    final myCounterText = find.text('-1');

    //Verifying widget using Matcher constant
    expect(myCounterText, findsOneWidget);
  });

  testWidgets('Incrementing & Decrementing Number',
      (WidgetTester widgetTester) async {
    Widget testWidget =
        MediaQuery(data: MediaQueryData(), child: CounterApp());
    await widgetTester.pumpWidget(testWidget);

    //Tap increment Button (Green)
    await widgetTester.tap(find.byKey(Key('I')));

    await widgetTester.pump();

    //Tap increment Button (Red)
    await widgetTester.tap(find.byKey(Key('D')));

    await widgetTester.pump();

    //Creating Finder to widget created
    final myCounterText = find.text('0');

    //Verifying widget using Matcher constant
    expect(myCounterText, findsOneWidget);
  });
});
}
```

SOURCE CODE ONLINE

The full source code for this example (Chapter 21: Widget Testing) is available at GitHub. The *CounterApp* code is available here (Chapter 21: Widget Testing).

RUNNING WIDGET TESTS

The widget tests can be run using the same command as unit tests.

```
$ flutter test test/main_21_test.dart
```

OUTPUT

The total time taken for executing all tests is displayed as `00:04`. The `+6` represents the number of tests executed. The last part, `All tests passed!`, displays the status of the test run.

```
00:04 +6: All tests passed!
```

CONCLUSION

In this chapter, you learned to write widget tests for *CounterApp*. You learned to use the `flutter _ test` package to write tests for verifying the behavior of widgets. You used WidgetTester, testWidgets(), Finder, and Matcher to test the widgets' behavior. The WidgetTester helped build interactive test widgets. The testWidgets()methods were used to write test code for verifying a widget's behavior. The Finder class is used to search for widgets in the test app. The Matcher class helped verify whether the Finder class could locate a given widget in the test environment.

REFERENCES

Dart Dev. (2020, 11 30). *test*. Retrieved from pub.dev: https://pub.dev/packages/test

Flutter Dev. (2020, 11 30). *Finder class*. Retrieved from api.flutter.dev: https://api.flutter.dev/flutter/flutter_test/Finder-class.html

Flutter Dev. (2020, 11 30). *Key class*. Retrieved from api.flutter.dev: https://api.flutter.dev/flutter/foundation/Key-class.html

Flutter Dev. (2020, 11 30). *Matcher class*. Retrieved from api.flutter.dev: https://api.flutter.dev/flutter/package-matcher_matcher/Matcher-class.html

Flutter Dev. (2020, 11 30). *MediaQuery class*. Retrieved from api.flutter.dev: https://api.flutter.dev/flutter/widgets/MediaQuery-class.html

Flutter Dev. (2020, 11 30). *pump method*. Retrieved from api.flutter.dev: https://api.flutter.dev/flutter/flutter_test/TestWidgetsFlutterBinding/pump.html

Flutter Dev. (2020, 11 30). *pumpWidget method*. Retrieved from flutter.dev: https://api.flutter. dev/flutter/flutter_test/WidgetTester/pumpWidget.html

Flutter Dev. (2020, 11 30). *tap method*. Retrieved from api.flutter.dev: https://api.flutter.dev/ flutter/flutter_test/WidgetController/tap.html

Flutter Dev. (2020, 11 30). *testWidgets function*. Retrieved from api.flutter.dev: https://api. flutter.dev/flutter/flutter_test/testWidgets.html

Flutter Dev. (2020, 11 30). *WidgetTester class*. Retrieved from api.flutter.dev: https://api. flutter.dev/flutter/flutter_test/WidgetTester-class.html

Flutter Team. (20202, 11 27). *flutter_test*. Retrieved from flutter.dev: https://flutter.dev/docs/ cookbook/testing/widget/introduction#1-add-the-flutter_test-dependency

Google. (2020, 11 30). *Widget Testing*. Retrieved from flutter.dev: https://flutter.dev/docs/ cookbook/testing/widget/introduction

Tyagi, P. (2020, 11 30). Chapter 21: *Widget Testing*. Retrieved from Pragmatic Flutter GitHub Repo: https://github.com/ptyagicodecamp/pragmatic_flutter/blob/master/test/ main_21_test.dart

Tyagi, P. (2021). Chapter 21: Unit Testing. In P. Tyagi, *Pragmatic Flutter: Building Cross-Platform Mobile Apps for Android, iOS, Web & Desktop*. CRC Press.

22 Integration Testing

In the previous chapter on Unit Testing (Chapter 20: Unit Testing), you learned to test the logic of functions, classes, and methods. Later, in the Widget Testing chapter (Chapter 21: Widget Testing) you learned to test widgets. In this chapter, you will learn to test an app on a real device while interacting with other system-level components. Integration testing allows running multiple pieces together. It's useful to test the performance of an app on real hardware.

Integration testing is done using two files known as 'test pair'. The first file deploys the instrumented application (Instrument the app) to the test device. It could either be a real device or an emulator. The second file contains the test cases that drive the application to execute the app's actions.

PACKAGE DEPENDENCY

The integration testing requires the testing pair of two files. The `flutter _ driver` (flutter_driver library) package needs to be included in the *pubspec.yaml* file. This package provides the tools to create the test pair. You need the following package added under the `dev _ dependencies` section of the *pubspec.yaml* config file.

```
```
dev_dependencies:
 flutter_test:
   sdk: flutter

 flutter_driver:
   sdk: flutter
```
```

PREVIEW *CounterApp* CODE

In this section, we will be writing integration tests for the *CounterApp* with two buttons. The application's code from the Unit Testing chapter (Chapter 20: Unit Testing) for your reference is as below:

```
```
void main() {
  runApp(CounterApp());
}

class CounterApp extends StatelessWidget {
  @override
  Widget build(BuildContext context) {
    return MaterialApp(
```
```

```
        debugShowCheckedModeBanner: false,
        theme: ThemeData(
          primarySwatch: Colors.blue,
          visualDensity: VisualDensity.adaptivePlatformDensity,
        ),
        home: MyHomePage(title: 'My Counter App'),
      );
    }
}

class MyHomePage extends StatefulWidget {
  MyHomePage({Key key, this.title}) : super(key: key);

  final String title;

  @override
  _MyHomePageState createState() => _MyHomePageState();
}

class Counter {
  int number = 0;

  void increment() => number++;
  void decrement() => number--;
}

class _MyHomePageState extends State<MyHomePage> {
  Counter _counter = Counter();

  void _incrementCounter() {
    setState(() {
      _counter.increment();
    });
  }

  void _decrementCounter() {
    setState(() {
      _counter.decrement();
    });
  }

  @override
  Widget build(BuildContext context) {
    return Scaffold(
      appBar: AppBar(
        title: Text(widget.title),
      ),
      body: Center(
        child: Column(
          mainAxisAlignment: MainAxisAlignment.spaceEvenly,
          children: <Widget>[
```

```
            MyTextWidget(
              myKey: Key('L'),
              textString: 'My Counter:',
              style: Theme.of(context).textTheme.headline4,
            ),
            MyTextWidget(
              myKey: Key('N'),
              textString: '${_counter.number}',
              style: Theme.of(context).textTheme.headline3,
            ),
            Row(
              mainAxisAlignment: MainAxisAlignment.
spaceEvenly,
              children: [
                ElevatedButton(
                  key: Key('D'),
                  style: ElevatedButton.styleFrom(primary:
Colors.red),
                  onPressed: () => _decrementCounter(),
                  child: FaIcon(FontAwesomeIcons.minus),
                ),
                ElevatedButton(
                  key: Key('I'),
                  style: ElevatedButton.styleFrom(primary:
Colors.green),
                  onPressed: () => _incrementCounter(),
                  child: FaIcon(FontAwesomeIcons.plus),
                )
              ],
            )
          ],
        ),
      ),
    );
  }
}


class MyTextWidget extends StatelessWidget {
  final String textString;
  final TextStyle style;
  final Key myKey;

  const MyTextWidget({this.myKey, this.textString, this.style});

  @override
  Widget build(BuildContext context) {
    return Text(
      textString,
      style: style,
      //Need to specify the directionality because this widget
will be tested independently.
```

```
    //You don't need to specify textDirection in your Text
widget when it's a child to Scaffold widget because
textDirection is implicitly provided in that case.
      textDirection: TextDirection.ltr,
      key: myKey,
    );
  }
}
```

TEST PAIR FILES

The test pair has two files, which are located in a directory named `test _ driver`.
This directory is located at the top-level of the Flutter project.

INSTRUMENTATION APP

The first file is created in the `test _ driver/` directory. You can name it as
you like. I named it that goes with the name of this chapter: `test _ driver/
main _ 22.dart`. This file is an instrumented version of the app and helps record
the instrumentation information like performance profiles to the integration tests
suite. The contents of this file look like below:

```
//Instrumented CounterApp
import 'package:flutter_driver/driver_extension.dart';
import '../lib/chapter22/main_22.dart' as app;

void main() {
  enableFlutterDriverExtension();
  app.main();
}
```

TEST SUITE

The second file is also present in the same directory, `test _ driver/`. It has the
same name as the instrumented file but suffixed with ` _ test`. So, the integration
test suites will be present in the *test_driver/main_22_test.dart* file. In the next section, we will write integration tests to test the increment and decrement buttons of
the *CounterApp*.

WRITING INTEGRATION TESTS

In this section, we will be writing test cases to test the app's functionality and user
interface interactions. There are three parts to writing integration test suites.

Test Suite

The multiple integration tests can be grouped based on their relevance in groups. In this section, we will create one group for testing increment and decrement functionality in *CounterApp*.

```
void main() {
    group('Description about the test suite', () {
    //Test cases go here
    });
}
```

Setting Up

In this test, we will be incrementing and decrementing the number by pressing the buttons and verifying the number displayed on the screen. We acquire a reference to each of these widgets using the `find.byValueKey()` method. The `numberTextFinder` stores a reference to the widget rendering the number. The `incrementFinder` holds the reference to the `ElevatedButton` widget for incrementing the counter. The `decrementFinder` holds the reference to the `ElevatedButton` widget for decrementing the counter.

```
final numberTextFinder = find.byValueKey('N');
final incrementFinder = find.byValueKey('I');
final decrementFinder = find.byValueKey('D');
```

Next, we'll get the reference to `FlutterDriver` (flutter_driver library), which connects the test suites to the instrumented app to verify the test cases.

```
FlutterDriver flutterDriver;
//connect to the instrumented app in test_driver/main_22.dart
setUpAll(() async {
  flutterDriver = await FlutterDriver.connect();
});
```

Tearing Down

The `tearDownAll()` function is called after all test cases are executed to close the app's connection.

```
//Close the connection between driver and instrumented app
tearDownAll(() async {
```

```
  if (flutterDriver != null) {
    flutterDriver.close();
  }
});
```

Test Case

In this test case, we will increment the counter number using the green button and decrement it twice. Increasing the number by one will render '1' on the screen. Decrementing it two times will render the number as '–1'.

The `ElevatedButton` widget to increment the number is tapped using `flutterDriver.tap(incrementFinder)`. Similarly, the button to decrement number is tapped using `flutterDriver.tap(decrementFinder)`. Tapping on either of the buttons will increase or decrease the number displayed. This number can be retrieved using `flutterDriver.getText(numberTextFinder)`. Finally, you can verify the expected behavior using `expect(flutterDriver. getText(numberTextFinder), 'number')`. In the following test case, the increment button is tapped once and verified as '1'. Next, the decrement button is tapped twice, which will decrease the number by two. Subtracting two from one gives '–1'. The '–1' is verified using `expect(await flutterDriver. getText(numberTextFinder), '-1');`.

```
test(' Increment & Decrement Counter', () async {
  //Tap increment counter
  await flutterDriver.tap(incrementFinder);

  //Verify if Number text becomes 1
  expect(await flutterDriver.getText(numberTextFinder), '1');

  //Tap decrement counter twice
  await flutterDriver.tap(decrementFinder);
  await flutterDriver.tap(decrementFinder);

  //Verify if Number text becomes -1
  expect(await flutterDriver.getText(numberTextFinder), '-1');
});
```

Complete Test Code

The complete test code for this integration test suite is as below:

```
//Integration Testing: CounterApp
import 'package:flutter_driver/flutter_driver.dart';
import 'package:test/test.dart';
```

```
void main() {
  group('Integration Testing CounterApp:', () {
    final numberTextFinder = find.byValueKey('N');
    final incrementFinder = find.byValueKey('I');
    final decrementFinder = find.byValueKey('D');

    FlutterDriver flutterDriver;

    //connect to the instrumented app in test_driver/main_26.dart
    setUpAll(() async {
      flutterDriver = await FlutterDriver.connect();
    });

    //Close the connection between driver and instrumented app
    tearDownAll(() async {
      if (flutterDriver != null) {
        flutterDriver.close();
      }
    });

    test(' Increment & Decrement Counter', () async {
      //Tap increment counter
      await flutterDriver.tap(incrementFinder);

      //Verify if Number text becomes 1
      expect(await flutterDriver.getText(numberTextFinder), '1');

      //Tap decrement counter twice
      await flutterDriver.tap(decrementFinder);
      await flutterDriver.tap(decrementFinder);

      //Verify if Number text becomes -1
      expect(await flutterDriver.getText(numberTextFinder),
'-1');
    });
  });
}
```

SOURCE CODE ONLINE

Source code for this integration test suite (Chapter 22: Integration Testing) is available at GitHub. The source code for companion *CounterApp* is available here (Chapter 22: Integration Testing).

RUNNING INTEGRATION TESTS

The integration tests are run using the following command on terminal (command line):

```
` ` `
$ flutter drive --target=test_driver/main_22.dart --browser-
name=chrome -debug
` ` `
```

CONCLUSION

In this chapter, you learned to setup and write a test suite for integration tests. The test suite helped test the *CounterApp* functionality end-to-end by increasing and decreasing the counter displayed on the screen and verifying the results.

REFERENCES

Flutter Dev. (2020, 11 30). *flutter_driver library*. Retrieved from api.flutter.dev: https://api. flutter.dev/flutter/flutter_driver/flutter_driver-library.html

Flutter Dev. (2020, 11 30). *Instrument the app*. Retrieved from Flutter Dev: https://flutter.dev/ docs/cookbook/testing/integration/introduction#4-instrument-the-app

Tyagi, P. (2020, 11 30). *Chapter 22: Integration Testing*. Retrieved from Pragmatic Flutter GitHub Repo: https://github.com/ptyagicodecamp/pragmatic_flutter/tree/master/test_ driver

Tyagi, P. (2020, 11 30). *Chapter 22: Integration Testing*. Retrieved from Pragmatic Flutter GitHub Repo: https://github.com/ptyagicodecamp/pragmatic_flutter/tree/master/lib/ chapter22

Tyagi, P. (2021). Chapter 20: Unit Testing. In P. Tyagi, *Pragmatic Flutter: Building Cross-Platform Mobile Apps for Android, iOS, Web & Desktop*. CRC Press.

Tyagi, P. (2021). Chapter 21: Widget Testing. In P. Tyagi, *Pragmatic Flutter: Building Cross-Platform Mobile Apps for Android, iOS, Web & Desktop*. CRC Press.

23 Rolling into the World

Congratulations! You made it to the end of the book. Now that your app is looking good in your development environment, you may want to share it with others. In this chapter, you will learn to release Android, iOS, web, and desktop applications built on the Flutter platform. The Android applications are distributed on Google Play Store (Google Play Store). Google Play is a digital distribution service for apps by Google. It distributes apps on Android and Chrome OS platforms. Chrome OS is based on Chromium OS (Chromium OS), which is an open-source project to build a fast, simple, and secure operating system targeted to web users. The Chrome OS (Chrome OS) is a Linux kernel-based operating system that uses Google Chrome web browser as its user interface. Google Chrome (Google Chrome) is a cross-platform web browser developed by Google.

In this chapter, you'll learn the basics of preparing your app for publishing to Google Play Store – Android, App Store Connect – iOS (Apps). You'll also learn to deploy the web app to Firebase hosting and building a self-contained desktop app that can be distributed to macOS users.

ADDING LAUNCHER ICON

An effective way to update app icons for Android and iOS variants is to use the *flutter_launcher_icons* (flutter_launcher_icons) package. This package helps to resize the icons for different screen form factors and devices accordingly and update icons for Android and iOS all at once. Let's see how to use it in a Flutter project.

ADDING DEPENDENCY

Add dependency in *pubspec.yaml*:

```
```
dependencies:
  flutter_launcher_icons: ^0.8.1
```
```

After adding a dependency in the *pubspec.yaml*, don't forget to fetch it using `flutter pub get` in the terminal.

CONFIGURING *pubspec.yaml*

Add your app icon in the *assets* directory of the Flutter project's root level. You can choose to organize launcher-related images in their own folder like *assets/launcher*. Android supports adaptive icons (Adaptive icons), and it requires two layers: foreground and background images to make an icon. You need a background image, additionally. You can choose to keep foreground and background images separately as two images or use a solid-colored image as the background and icon as the foreground image. Assume that the app icon image is *assets/launcher/app_logo.png*. The background layer image is *assets/launcher/app_logo_background.png*.

```
```
flutter_icons:
 ios: true
 android: true
 image_path_ios: "assets/launcher/app_logo.png"
 image_path_android: "assets/launcher/app_logo.png"
 adaptive_icon_background: "assets/launcher/app_logo_
background.png"
 adaptive_icon_foreground: "assets/launcher/app_logo.png"
```
```

GENERATING LAUNCHER ICONS

Running the following command in the terminal will generate the required app icons for Android and iOS platforms.

```
```
flutter pub run flutter_launcher_icons:main
```
```

The above command will generate icons in '*android/app/src/main/res/mipmap-**' for the Android platform. The adaptive icons are generated in '*android/app/src/main/res/drawable-**' folders.

Icons for iOS are generated/replaced in the '*ios/Runner/Assets.xcassets/*' folders. The desktop macOS app icons can be updated in '*macos/Runner/Assets.xcassets/AppIcon.appiconset/*' folder. The web icons need to be updated in the '*web/icons/*' folder.

RELEASING ANDROID APPS

In this section, you'll get pointers on building and preparing your application to distribute on Google Play Store (Play Store). If this is your first-time publishing apps on Play Store, then you would need to pay a $25 developer registration fee for signing up (Create a new developer account) on the publishing platform.

An app needs to be digitally signed to be able to be published on Google Play Store. There are two types of artifacts supported by the Google Play Store: App Bundle and APK.

Note: Make sure to run `flutter analyze` before you build artifacts for publishing to Play Store.

CREATE KEYSTORE

You create a keystore only once for the application's lifetime in the store. Use the following command to generate a keystore in the following environments. The `keystore.jks` is the keystore file and meant to be kept private.

```
```
keytool -genkey -v -keystore <path-to-dir>/keystore.jks
-keyalg RSA -keysize 2048 -validity 10000 -alias key
```
```

The above commands will ask a couple of questions related to your organization that you will provide in the command line. You'll also be providing a password. Once you provide all the required information, you'll have *keystore.jks* generated in the given directory. In case you're planning to keep this keystore file in your project directory, make sure to add it to the '*.gitignore*' file to avoid accidental check-in to the public repository. Learn more about creating keystores here (Create a keystore).

Note: The keytool tool comes with Java installation. Run `flutter doctor -v` to find Java binary installation that came with Android Studio.

REFERENCING KEYSTORE

Add a file `key.properties` in the `<flutter-project-root>/android` folder to reference `keystore.jks` and other related details. Add the `key.properties` file in the `.gitignore` as well. You don't want to check-in this file in the version control.

```
```
storePassword=<password created during keystore.jks
generation>
keyPassword=<password created during keystore.jks generation>
keyAlias=<alias chosen during keystore.jks generation>
storeFile=<path-to-keystore>/keystore.jks>
```
```

SIGNING CONFIGURATION

Android application's signing details are configured in `<flutter-project-root>/android/app/build.gradle` file.

LOADING `key.properties`

The configuration details are loaded from `key.properties` above the `android {}` block of the gradle file.

```
```
def keystoreProperties = new Properties()
def keystorePropertiesFile = rootProject.file('key.
properties')
if (keystorePropertiesFile.exists()) {
    keystoreProperties.load(new
FileInputStream(keystorePropertiesFile))
}
```
```

```
android {
...
}
```

ADDING RELEASE SIGNING CONFIGURATION

```
android {
    ...

    signingConfigs {
        release {
            keyAlias keystoreProperties['keyAlias']
            keyPassword keystoreProperties['keyPassword']
            storeFile keystoreProperties['storeFile'] ?
file(keystoreProperties['storeFile']) : null
            storePassword keystoreProperties['storePassword']
        }
    }
    buildTypes {
        release {
            signingConfig signingConfigs.release
        }
        debug {
            signingConfig signingConfigs.debug
        }
    }
}
```

Note: Run the `flutter clean` command to refresh the gradle configuration.

CODE SHRINKING

Add `minifyEnabled` and `shrinkResources` to true for R8 code shrinker (Shrink, obfuscate, and optimize) to automatically remove the code that's not required at runtime.

```
buildTypes {
    release {
        minifyEnabled true
        shrinkResources true
        signingConfig signingConfigs.release
    }
}
```

The R8 code shrinker is enabled by default when building a release artifact. However, it can be disabled by passing the `-no-shrink` flag while building release artifacts.

AndroidManifest.xml

Make sure that you've internet permission enabled in `<flutter-project-root>/android/app/src/main/AndroidManifest.xml`. It allows internet access during development to enable communication between Flutter tools and the app.

```
<uses-permission android:name="android.permission.INTERNET"/>
```

Another thing you want to ensure the name of your app is declared using the `android:label` attribute.

```
<application
    ...
    android:label="appName">
    ...
</application>
```

This is defined in the `name` attribute in the `pubspec.yaml`.

```
name: appName
```

BUILD CONFIGURATION

The app's default build configuration is declared in the `defaultConfig` block of `<flutter-project-root>/android/app/build.gradle`.

applicationId:
The `applicationId` contains the unique app id for your app. It can't be changed once an app is published on the Play Store.

```
defaultConfig {
    applicationId "com.domain.appId"
    minSdkVersion 18
    targetSdkVersion 28
    versionCode flutterVersionCode.toInteger()
    versionName flutterVersionName
    multiDexEnabled true
}
```

minSdkVersion:
Minimum Android API level supported by the app.

targetSdkVersion:
The targeted Android API level on which the app is designed to run.

versionCode & versionName:

This can be set in the `version` attribute in the `pubspec.yaml` file. The versionCode specifies the internal app version, while versionName specifies the
display string for the version in Play Store. The `version` in `pubspec.yaml` is
specified as `1.0.0+1`. The first part, `1.0.0` is the versionName, and everything after the `+` sign is the versionCode. You can update the versionName to
'1.0.1', and versionCode to '2' at once in pubspec.yaml file like below:

```
version: 1.0.1+2
```

BUILDING APP BUNDLE

The app bundles are the newer and recommended artifact type. Execute the following command in the Flutter project's root directory to build the release artifact in the
'*.aab' file format. It's generated in the `<flutter-project-root>/build/app/
outputs/bundle/release/` directory. The artifact created is `app-release.
aab`.

```
flutter build appbundle
```

The '*app-release.aab*' contains Dart and Flutter code compiled for ARM 32-bit
(armeabi-v7a), ARM 64-bit (arm64-v8a), and x86-64. There's only one artifact for
each target application binary interface (ABI).

BUILDING APK

An APK artifact is needed for the distribution platforms that don't support app
bundle formats. By default, one APK contains native binaries for all the platforms/
ABI(s) in one fat artifact and risks users to download files that are not an application
to their device. In this case, it makes sense to create an APK targeted to a specific
platform. The APK artifacts targeted to each application binary interface or ABI can
be created using the following command:

```
flutter build apk --split-per-abi
```

The above command will generate artifacts in multiple ABI targeted artifacts
in `<flutter-project-root>/build/app/outputs/flutter-apk/` and
`<flutter-project-root>/build/app/outputs/apk/release` directories. The artifacts generated in both directories are:

- app-arm64-v8a-release.apk
- app-armeabi-v7a-release.apk
- app-x86_64-release.apk

Distributing App on Google Play Store

In this section, we'll go over the basic steps to publish your app bundle '*.aab' artifact to the Play Store. I recommend reviewing the detailed launch checklist here (Google Play).

Creating Application

Once you're at Google Play's publishing portal (Google Play Console), click the 'Create Application' button to get started with publishing your app. You'll get a prompt to enter the name of your application. Once you enter the name for your app and choose the target language, you're redirected to the store listing next. Make sure to keep saving the draft as you make progress.

App Releases

In this section of the portal, you upload the '*.aab' artifact. There are four tracks that you can choose to upload your app bundle/APK artifacts.

1. Production track (Production): In this track, the app becomes available to all users on the Google Play Store.
2. Open track (Beta): In this track, the app is available for open testing. Anyone with the link can access the app.
3. Closed track (Alpha): In this track, the app becomes available for closed testing. The app becomes available to allowed listed users.
4. Internal test track: In this track, the app becomes available for internal testing quickly. In contrast, other options may take several hours to become available. It requires the allowed testers listed.

It's recommended to upload artifacts for internal testing first and then gradually move up to the production track. This strategy helps to catch the bugs and issues much earlier to provide a quality experience to users worldwide later.

Store Listing

This is where you provide the app's short and detailed description. Be ready with a high-resolution app icon (512 × 512), feature graphic (1024 w × 500 h), and screenshots for supported devices like phones and tablets in our case. You can use the Android Asset Studio (Launcher icon generator) to generate app icons. All screenshot images' sizes should follow a 2:1 ratio.

Choose the application type and category for the app. It's required to apply a content rating on your app listing. Take a content rating questionnaire to apply a content rating. You need to upload your '*.aab' artifact before you can take a content rating questionnaire. Provide your contact details under the 'Contact details' part of the Store Listing.

App Content

In this section, you need to provide details about the privacy policy, ads integration, granular login-based app access, target audience information, and content rating.

Price & Distribution

In this section, the price and availability of the app are declared. Information about the content guidelines and US export laws is found in this section.

At this point, the app is ready to be published. There are many other options to customize your app's listing that you may want to go over in the Play Store publishing portal.

RELEASING iOS APPS

This section will touch base briefly on preparing and distributing iOS apps on App Store Connect. The detailed instructions on building and releasing an iOS app are available at Flutter's official documentation (Build and release an iOS app).

Your app must meet the App Store's review guidelines (App Review). If this is your first-time publishing iOS apps, then you need to enroll in the Apple Developer Program (Develop) and choose one of the available membership programs (Choosing a Membership).

Register Identifier (bundle ID)

An iOS app has a Bundle ID. It's a unique identifier registered on Apple Developer. Create Bundle ID (Register a new identifier) on your Apple Developers account. Select 'App' and follow the directions, as shown in Figure 23.1. Choose 'Explicit App ID' when the option is available.

Register App

Register your app at App Store Connect (App Store Connect). Open the App Store Connect in a browser and choose the 'My Apps' icon. Click on the '+' button next to 'Apps' as shown in Figure 23.2.

You will get a dialog, as shown in Figure 23.3, to enter details about your new app. Once you enter the app's basic details, click on 'Create' to add the app. You'll be

FIGURE 23.1 Registering a new identifier

FIGURE 23.2   App Store Connect: Add new app

presented with a page to add detailed information about the app. Check out official documentation (Add a new app) on adding an app to the App Store Connect.

XCODE PROJECT SETTINGS

Open the Flutter's iOS variant in Xcode to review (Review Xcode project settings) the settings. First, open Xcode IDE and choose the 'Open a project or file' option and navigate to '*ios/Runner.xcworkspace*'. This will open the project in Xcode.

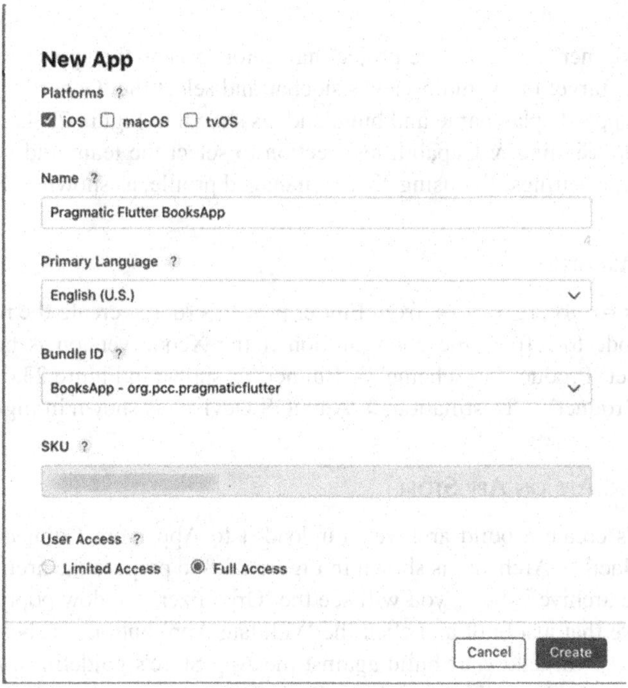

New App

Platforms ?

☑ iOS ☐ macOS ☐ tvOS

Name ?

Pragmatic Flutter BooksApp

Primary Language ?

English (U.S.)   ∨

Bundle ID ?

BooksApp - org.pcc.pragmaticflutter   ∨

SKU ?

User Access ?

○ Limited Access   ● Full Access

Cancel   Create

FIGURE 23.3   App Store Connect: New app details

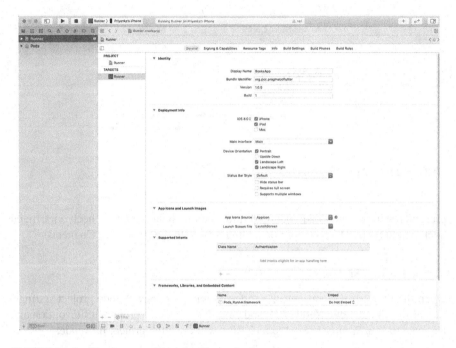

FIGURE 23.4   Xcode project settings: General tab

Select 'Runner' in the Xcode project navigator to view the app's settings. Select the 'Runner' target in the main view's sidebar and select the 'General' tab. You can update the app's display name and bundle id, as shown in Figure 23.4.

Select the 'Signing & Capabilities' section to select the team and update provisioning profile settings. I'm using Xcode managed profile, as shown in Figure 23.5.

BUILDING ARCHIVE

Run `flutter build ios` from Flutter project's root to create the release build. Restart Xcode to refresh the configuration if the Xcode version is below 8.3. In Xcode, select 'Product' > 'Scheme' > 'Runner' as shown in Figure 23.6.

Select 'Product' > 'Destination' > 'Any iOS Device' as shown in Figure 23.7.

DISTRIBUTING APP ON APP STORE

Finally, let's create a build archive to upload it to App Store Connect. In Xcode, select 'Product' > 'Archive' as shown in Figure 23.8 to prepare the archive.

Once the archive is built, you will see the 'Organizer' window popup. Select the build archive that just built and click the 'Validate App' button, as shown in Figure 23.9. This will validate your build against the App Store's guidelines. Address any reported issues and rebuild the archive. Once the archive is successfully built, click on 'Distribute App' button, as shown in Figure 23.9.

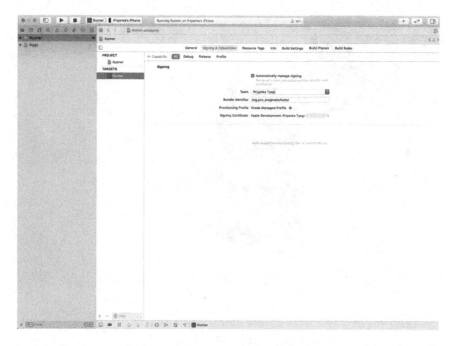

FIGURE 23.5   Xcode project settings: Signing & Capabilities tab

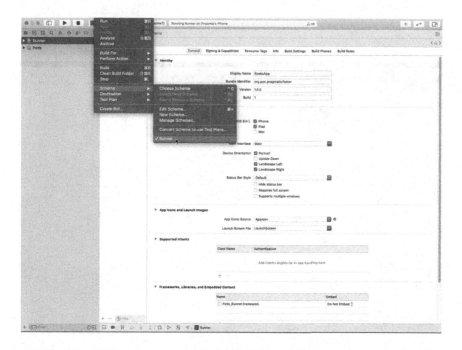

FIGURE 23.6   Xcode project settings: Product > Scheme > Runner

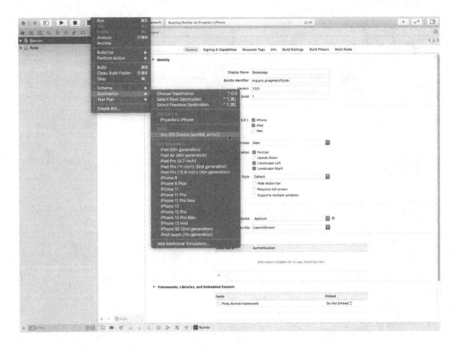

FIGURE 23.7   Xcode project settings: Product > Destination > Any iOS Device

FIGURE 23.8   Xcode project settings: Product > Archive

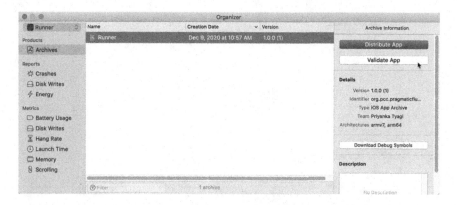

FIGURE 23.9 Organizer: Validate App & Distribute App

You will see a dialog to choose the method of distribution. Choose 'App Store Connect', as shown in Figure 23.10 and click the 'Next' button.

Select the 'Upload' option in Figure 23.11 and follow the screens to upload archive (Upload an app to App Store Connect) to App Store Connect.

Once your build is uploaded to App Store Connect, prepare it for submission, and set the pricing and availability details. Lastly, click on 'Submit for Review'. You should get notified from Apple about the app's review status updates in 24–48 hours.

RELEASING WEB APPS

The Flutter Web provides support for compiling Flutter's Dart code into native client code that can be deployed to any web server of your choice. I will give pointers to setup the Firebase project for hosting web app BooksApp created in the previous chapter (Chapter 15: The Second Page: BookDetails Widget).

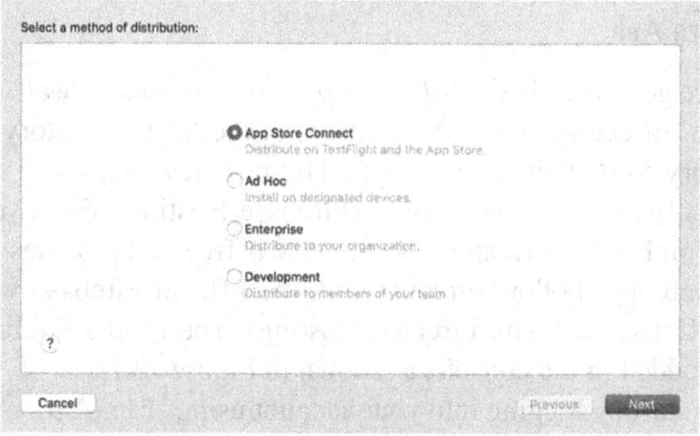

FIGURE 23.10 Selecting a method of distribution

FIGURE 23.11   Selecting a destination

Building WebApp

Run the following command to create a release build for a web app.

```
flutter build web
```

The above command generates web content in the '*build/web*' folder. Everything in this folder needs to be moved to the hosting folder. A Flutter web app release build can also be created using the alternative command as below:

```
flutter run --release
```

The above commands use the dart2js (dart2js: Dart-to-JavaScript compiler) compiler to produce '*main.dart.js*' JavaScript file for '*main.dart*'.

Deploying Web App

The artifacts are generated in the '*<Flutter-project-root>/build/web*' directory along with the '*assets*' directory. Move the contents of the '*web*' directory to the publicly hosting directory to deploy the web app. The next few steps will describe how I deployed *BooksApp* to Firebase hosting (Firebase Hosting). First, you need to create an account on Firebase Hosting. Its first tier is free and provides enough tooling to host your web app. Follow directions on the official Firebase website to setup Firebase CLI (Get started with Firebase Hosting). You need a Firebase project created and a site added in the console, as shown in Figure 23.12.

You need to start by logging into your account using `firebase login`. Once you're logged in, run `firebase init` to setup the hosting. This will create a few configuration files like *firebase.json* and *.firebaserc*.

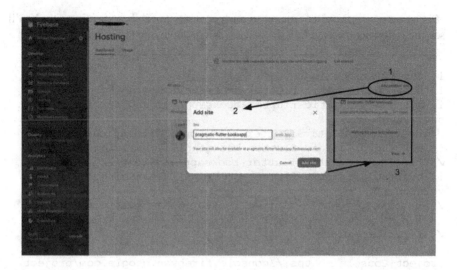

FIGURE 23.12   Adding site name to Firebase hosting portal

firebase.json:

```
~~~
{
  "hosting": {
    "site": "pragmatic-flutter-booksapp",
    "public": "public/web",
    "ignore": [
      "firebase.json",
      "**/.*",
      "**/node_modules/**"
    ]
  }
}
~~~
```

.firebaserc:

```
~~~
{
  "projects": {
    "default": "firbase_project_id"
  }
}
~~~
```

I prefer to create a 'deploy' folder (Chapter 23: Rolling into the World) in the root of the project to save web hosting related content. Check it out at GitHub repo.

In the *deploy* folder, *'public/web'* contains hosting contents. You need to move contents from the *'build/web'* to the *'deploy/public'* folder. Finally, run the `firebase deploy` command to push contents to Firebase hosting site. You'll see the following messages on the console while deploying to Firebase:

```
~~~
i  deploying hosting
i  hosting[pragmatic-flutter-booksapp]: beginning deploy...
i  hosting[pragmatic-flutter-booksapp]: found 52 files in
public/web
✓  hosting[pragmatic-flutter-booksapp]: file upload complete
i  hosting[pragmatic-flutter-booksapp]: finalizing version...
✓  hosting[pragmatic-flutter-booksapp]: version finalized
i  hosting[pragmatic-flutter-booksapp]: releasing new
version...
✓  hosting[pragmatic-flutter-booksapp]: release complete

✓  Deploy complete!

Project Console: https://console.firebase.google.com/project/
fir-recipes-b5611/overview
Hosting URL: https://pragmatic-flutter-booksapp.web.app
~~~
```

The *BooksApp* web app is available on the hosting URL now (BooksApp Flutter Web App).

RELEASING DESKTOP (macOS) APPS

At the time of this writing (Distribution), desktop support is available in the Flutter dev channel (Flutter's channels - dev). It's not recommended to release a desktop application until its support becomes stable.

SETTING UP ENTITLEMENTS

Setting up entitlements (macOS-specific support) are necessary to access device capabilities. You need to add sandbox entitlement to enable the app to distribute to the App Store. The network entitlement is needed to make network calls to connect to the Books API to fetch data. These entitlements need to be added to the '*macos/ Runner/Release.entitlements*' file.

```
~~~
<?xml version="1.0" encoding="UTF-8"?>
<!DOCTYPE plist PUBLIC "-//Apple//DTD PLIST 1.0//EN"
"http://www.apple.com/DTDs/PropertyList-1.0.dtd">
<plist version="1.0">
<dict>
    <key>com.apple.security.app-sandbox</key>
    <true/>
    <key>com.apple.security.network.client</key>
     <true/>
</dict>
```

```
</plist>
```
` ` `

BUILDING RELEASE APP

Run the following command to create a release build for a macOS desktop app.

` ` `

```
flutter build macos
```
` ` `

The above command builds a self-contained application file with the suffix '.app' in the Flutter project's root in the *build* folder. The full path for the artifact for BooksApp is located at '*build/macos/Build/Products/Release/pragmatic_flutter.app*'.

DISTRIBUTING APP

There are two ways to distribute your macOS desktop application. You can distribute it on App Store (Distributing software on macOS). Distributing on the App Store makes it easier for users to discover and manage apps and helps you to send new updates to the app seamlessly. Another way is to distribute self-contained '*.app' to macOS users directly.

CONCLUSION

In this chapter, you learned the basics for releasing a Flutter application's Android, iOS, Web, and Desktop variants on to their respective distribution platforms. You learned to build release artifacts, preparing them for distribution, and releasing to the world.

REFERENCES

Android Developer. (2020, 12 09). *Google Play*. Retrieved from developer.android.com: https://developer.android.com/distribute/best-practices/launch

Android Developer. (2020, 12 09). *Shrink, obfuscate, and optimize*. Retrieved from developer.android.com: https://developer.android.com/studio/build/shrink-code

Apple Inc. (2020, 12 07). *Apps*. Retrieved from App Store Connect: https://appstoreconnect.apple.com/apps

Apple Inc. (2020, 12 08). *Distributing software on macOS*. Retrieved from developer.apple.com: https://developer.apple.com/macos/distribution/

Apple Inc. (2020, 12 09). *Add a new app*. Retrieved from help.apple.com: https://help.apple.com/app-store-connect/#/dev2cd126805

Apple Inc. (2020, 12 09). *App Review*. Retrieved from developer.apple.com: https://developer.apple.com/app-store/review/

Apple Inc. (2020, 12 09). *App Store Connect*. Retrieved from My Apps: https://appstoreconnect.apple.com/

Apple Inc. (2020, 12 09). *Choosing a Membership*. Retrieved from developer.apple.com: https://developer.apple.com/support/compare-memberships/

Apple Inc. (2020, 12 09). *Develop*. Retrieved from developer.apple.com: https://developer.apple.com/develop/

Apple Inc. (2020, 12 09). *Register a new identifier.* Retrieved from developer.apple.com: https://developer.apple.com/account/resources/identifiers/add/appId/bundle

Apple Inc. (2020, 12 09). *Upload an app to App Store Connect.* Retrieved from help.apple.com: https://help.apple.com/xcode/mac/current/#/dev442d7f2ca

Dart Dev. (2020, 12 08). *dart2js: Dart-to-JavaScript compiler.* Retrieved from dart.dev: https://dart.dev/tools/dart2js

Firebase. (2020, 12 08). *Get started with Firebase Hosting.* Retrieved from firebase.google.com: https://firebase.google.com/docs/hosting/quickstart

Firebase Hosting. (2020, 12 08). Retrieved from console.firebase.google.com: https://firebase.google.com/products/hosting

Flutter Community. (2020, 12 07). *flutter_launcher_icons.* Retrieved from pub.dev: https://pub.dev/packages/flutter_launcher_icons

Flutter Dev. (2020, 12 08). *Distribution.* Retrieved from Desktop support for Flutter: https://flutter.dev/desktop#distribution

Flutter Dev. (2020, 12 08). *macOS-specific support.* Retrieved from Setting up entitlements: https://flutter.dev/desktop#setting-up-entitlements

Flutter Dev. (2020, 12 09). *Build and release an iOS app.* Retrieved from flutter.dev: https://flutter.dev/docs/deployment/ios

Flutter Dev. (2020, 12 09). *Create a keystore.* Retrieved from flutter.dev: https://flutter.dev/docs/deployment/android#create-a-keystore

Flutter Dev. (2020, 12 09). *Review Xcode project settings.* Retrieved from flutter.dev: https://flutter.dev/docs/deployment/ios#review-xcode-project-settings

Flutter's channels - dev. (2020, 12 08). Retrieved from Flutter build release channels: https://github.com/flutter/flutter/wiki/Flutter-build-release-channels#dev

Google. (2020, 12 07). *Adaptive icons.* Retrieved from developer.android.com: https://developer.android.com/guide/practices/ui_guidelines/icon_design_adaptive

Google. (2020, 12 07). *Chromium OS.* Retrieved from The Chromium Projects: https://www.chromium.org/chromium-os

Google. (2020, 12 09). *Create a new developer account.* Retrieved from play.google.com: https://play.google.com/console/signup

Google. (2020, 12 09). *Google Play Console.* Retrieved from play.google.com: https://play.google.com/apps/publish

Google. (2020, 12 09). *Play Store.* Retrieved from play.google.com: https://play.google.com/store?hl=en_US&gl=US

Google Play Store. (2020, 12 07). *Google Play Store.* Retrieved from Google Play Store: https://play.google.com/store

Launcher icon generator. (2020, 12 09). Retrieved from romannurik.github.io: https://romannurik.github.io/AndroidAssetStudio/icons-launcher.html#foreground.type=clipart&foreground.clipart=android&foreground.space.trim=1&foreground.space.pad=0.25&foreColor=rgba(96%2C%20125%2C%20139%2C%200)&backColor=rgb(68%2C%20138%2C%20255)&crop=0

Tyagi, P. (2020, 12 08). *BooksApp Flutter Web App.* Retrieved from Firebase Hosting: https://pragmatic-flutter-booksapp.web.app/#/

Tyagi, P. (2020, 12 08). *Chapter 23: Rolling into the World.* Retrieved from Pragmatic Flutter GitHub Repo: https://github.com/ptyagicodecamp/pragmatic_flutter/tree/master/deploy

Tyagi, P. (2021). Chapter 15: The Second Page: BookDetails Widget. In P. Tyagi, *Pragmatic Flutter: Building Cross-Platform Mobile Apps for Android, iOS, Web & Desktop.* CRC Press.

Wikipedia. (2020, 12 07). *Chrome OS.* Retrieved from Wikipedia: https://en.wikipedia.org/wiki/Chrome_OS

Wikipedia. (2020, 12 07). *Google Chrome.* Retrieved from Wikipedia: https://en.wikipedia.org/wiki/Google_Chrome

Index

Printed in the United States
by Baker & Taylor Publisher Services